T0139202

XML in Scientific Computing

CHAPMAN & HALL/CRC
Numerical Analysis and Scientific Computing

Aims and scope:
Scientific computing and numerical analysis provide invaluable tools for the sciences and engineering. This series aims to capture new developments and summarize state-of-the-art methods over the whole spectrum of these fields. It will include a broad range of textbooks, monographs, and handbooks. Volumes in theory, including discretisation techniques, numerical algorithms, multiscale techniques, parallel and distributed algorithms, as well as applications of these methods in multi-disciplinary fields, are welcome. The inclusion of concrete real-world examples is highly encouraged. This series is meant to appeal to students and researchers in mathematics, engineering, and computational science.

Proposals for the series should be submitted to one of the series editors above or directly to:
CRC Press, Taylor & Francis Group
4th, Floor, Albert House
1-4 Singer Street
London EC2A 4BQ
UK

Published Titles

Classical and Modern Numerical Analysis: Theory, Methods and Practice
Azmy S. Ackleh, Edward James Allen, Ralph Baker Kearfott, and Padmanabhan Seshaiyer

Cloud-Computing: Data-Intensive Computing and Scheduling
Frédéric Magoulès, Jie Pan, and Fei Teng

Computational Fluid Dynamics
Frédéric Magoulès

A Concise Introduction to Image Processing using C++
Meiqing Wang and Choi-Hong Lai

Decomposition Methods for Differential Equations: Theory and Applications
Juergen Geiser

Desktop Grid Computing
Christophe Cérin and Gilles Fedak

Discrete Variational Derivative Method: A Structure-Preserving Numerical Method for Partial Differential Equations
Daisuke Furihata and Takayasu Matsuo

Grid Resource Management: Toward Virtual and Services Compliant Grid Computing
Frédéric Magoulès, Thi-Mai-Huong Nguyen, and Lei Yu

Fundamentals of Grid Computing: Theory, Algorithms and Technologies
Frédéric Magoulès

Handbook of Sinc Numerical Methods
Frank Stenger

Introduction to Grid Computing
Frédéric Magoulès, Jie Pan, Kiat-An Tan, and Abhinit Kumar

Iterative Splitting Methods for Differential Equations
Juergen Geiser

Mathematical Objects in C++: Computational Tools in a Unified Object-Oriented Approach
Yair Shapira

Numerical Linear Approximation in C
Nabih N. Abdelmalek and William A. Malek

Numerical Techniques for Direct and Large-Eddy Simulations
Xi Jiang and Choi-Hong Lai

Parallel Algorithms
Henri Casanova, Arnaud Legrand, and Yves Robert

Parallel Iterative Algorithms: From Sequential to Grid Computing
Jacques M. Bahi, Sylvain Contassot-Vivier, and Raphael Couturier

Particle Swarm Optimisation: Classical and Quantum Perspectives
Jun Sun, Choi-Hong Lai, and Xiao-Jun Wu

XML in Scientific Computing
C. Pozrikidis

XML in Scientific Computing

C. Pozrikidis

CRC Press is an imprint of the
Taylor & Francis Group, an **informa** business
A CHAPMAN & HALL BOOK

CRC Press
Taylor & Francis Group
6000 Broken Sound Parkway NW, Suite 300
Boca Raton, FL 33487-2742

Printed in the United States of America on acid-free paper
Version Date: 20120813

International Standard Book Number: 978-1-4665-1227-6 (Hardback)

Visit the Taylor & Francis Web site at
http://www.taylorandfrancis.com

and the CRC Press Web site at
http://www.crcpress.com

Contents

Preface

Xml stands for extensible markup language. In fact, *xml* is not a language, but a systematic way of encoding and formatting data and statements contained in an electronic file according to a chosen tagging system. A tag may represent a general entity, a physical, mathematical, or abstract object, an instruction, or a computer language construct. The data can describe cars and trucks in a dealer's lot, the chapters of a book, the input or output of a scientific experiment or calculation, the eigenvalues of a matrix, and anything else that can be described by numbers and words.

Data presentation and description

In the *xml* framework, information is described and presented in the same document, thus circumventing the need for legends and explanations. For example, we may order:

```
<breakfast> toast and eggs </breakfast>
```

Further cooking instructions can be included between the breakfast tag enclosed by the pointy brackets (<>) and its closure denoted by the slash (/).

Data reuse

Xml data (input) can be read by a person or parsed and processed by a program (application) that produces a new set of data (output). Although the input is the same, the output depends on the interpretation of the tags formatting the data. The inherent polymorphism allows us to materialize the same original data in different ways. For example:

1. An author may write a book inserting formatting tags between words, equations, and figures according to *xml* conventions and grammar. The text (data) file can be processed to produce books with different appearances.

2. A scientist may write a finite-element code that produces output tagged according to *xml* conventions. The elements can be visualized using different graphics programs and the data can be sent to another person or program to serve as input.

3. A conversation could be transcribed using *xml* grammar and then printed on paper or sent to a telephone to be heard by the recipient. It is not necessary to duplicate the data.

4. A computer program could be written according to generic *xml* conventions. The instructions can be interpreted to produce corresponding code in a chosen programming language.

To demonstrate the concept of data sharing and reuse, we deliver the same instructions to a painter and a sculptor, and ask them to produce corresponding pieces of art. The *xml* data encapsulated in these instructions acquire meaning only when the tags describing the data are implemented by the artists to produce physical objects.

Scientific computing

In scientific computing, we are accustomed to compiling and running a code (application) written in a language of our choice, such as C, C++, *fortran*, or *Matlab*®. The code utilizes parameters and input data that are either embedded in the program (monolithic structure) or read from companion input data files (modular structure). Emphasis is placed on the code and the output is generated readily by running the executable. In most applications, the code is more valuable than the output. The opposite is generally true in the *xml* framework where the data play a prominent role and may even serve to launch an application, as in the case of a telephone that rings only when it receives data.

Xml and scientific computing

Xml has received a great deal of attention in the web programming and software engineering disciplines with reference to data encoding and storage, but far less attention in the mainstream computational science and engineering disciplines. Two main issues of interest in scientific computing are: (*a*) producing *xml* formatted output from code and (*b*) reading *xml* input from a data file, converting it into an appropriate data structure. It is revealing that computing environments familiar to scientists and engineers, such as *Matlab*®, *Mathcad*®, and *Mathematica*®, have embraced the *xml* framework and incorporated add-on libraries to facilitate the handling of *xml* input and output.

Goals of this book

Currently available texts and *web* tutorials on *xml* data formatting discuss *xml* in the context of computer science with a clear focus on *web* and database programming.

The first goal of this book is to introduce and describe *xml* to scientists and engineers with some typesetting and programming experience.

The second goal is to introduce the extensible stylesheet language (*xsl*) with applications in *xml* data processing and numerical computation. Strange though it may seem, an *xsl* code is written according to *xml* conventions, that is, *xsl* is an *xml* implementation.

The third and perhaps most important goal of this book is to review possible ways of saving, importing, and sharing *xml* data in code written in programming languages used most frequently by scientists and engineers. Although references to *latex*, *html*, *fortran* 77 (simply called *fortran*), C++, and *perl* are made, only cursory familiarity with these languages is assumed and necessary explanations are given. Analogies and parallels will be drawn, and contrasts will be made with *xsl* to underline important similarities and differences in programming procedures.

This book is accompanied by a suite of computer programs and other documents arranged in directories corresponding to the book chapters and appendices.* Internet resources and other information pertinent to *xml* are provided as links at the book website.

C. Pozrikidis

Summer, 2012

---

*http://dehesa.freeshell.org/XML

Notation

Nomenclature and font conventions adopted in the text are defined in the following table:

Symbol or word	Name or meaning
()	parentheses
[]	square brackets
{}	curly brackets
<>	angle (pointy) brackets
->	*ascii* arrow
filename	name of a file
`sometext`	text typed in a file
language	name of a computer language
`line`	text typed in the keyboard
result	text shown in the screen
ENTER	Enter key in the keyboard

The names of standard computer languages are treated as regular words whose initial letter is capitalized at the beginning of a sentence and printed in lower case otherwise. File contents, typed instructions, and other data appearing on a computer screen are highlighted.

1 *Text and data formatting*

An attractive document contains well structured and sensibly formatted sentences, paragraphs, tables, and other elements. To create a document in a computer, we type words, characters, punctuation and other symbols in an electronic file, and then insert *tags* implementing formatting. In practice, most documents are created using advanced word editing programs (applications). Formatting is typically enforced by selecting (highlighting) text with the computer mouse and then clicking on embellishing buttons to underline, set in bold face, change the font size, or implement a wide range of other features offered in menus. The application automatically generates formatting tags that are implicitly contained in the document but are not displayed on the computer screen, unless optionally requested.

Different applications employ different tagging systems for implementing formatting instructions. A tagging system is also known as a *markup language.* The characterization of a tagging system as a *language* is not entirely appropriate. A computer language conveys instructions that implement deliberate procedures, such as algorithms, whereas a spoken language conveys content (data) and form (markup). Although unformatted data are useful, markup without data comprises an empty form. Nevertheless, the terminology *markup language* is broadly accepted to indicate a tagging system as standard semantics.

Reviewing two popular markup languages, *latex* and *html*, naturally leads us to the concept of the extensible markup language (*xml*), where data are presented *and* described using tags of our choice in a self-contained document. The information contained in an *xml* file can be read by a person or processed by a computer application written in a computer language of choice. The data can be parsed and modified in some desirable fashion, or else imported into a computer code for use in professional, scientific, and engineering applications. In fact, we will see that *xml* provides us with a framework for implementing computer instructions and formatting programming components. The salient features of *xml* and its relevance in scientific computing will be reviewed in this chapter.

1.1 Text formatting with latex and html

Two important markup languages (tagging systems) are the *latex* markup language used for typesetting documents, and the hypertext markup language (*html*) used for displaying documents in an Internet browser.* *Latex* was conceived by computer science Professor Donald Knuth and first released in 1978 with scientific document preparation in mind. *Html* was developed at the dawn of the Internet in the late 1980s with electronic document communication in mind. In spite of differences in their intended usage, the two languages have a similar structure and a parallel design. Contemplating the similarities and differences between these two languages naturally leads us to the concept of *xml* as a generalized framework.

1.1.1 Latex

A *latex* tag is a keyword preceded by the backslash (\). Square and curly brackets following the keyword enclose data and parameters. A perfectly valid complete *latex* file entitled *sample.tex* may contain the following lines:

```
\documentclass[11pt, letter]{article}
\begin{document}

    \textit{This sentence is set in italic}

\end{document}
```

The first line defines the size of the standard font (eleven points) to be used in the document, the dimensions of the paper where the text will be printed (letter-sized paper), and the type of document to be produced (article). If we were writing a book, we would have entered *book* instead of *article* in the first line. Document classes are available for scientific papers and seminar presentations.

The beginning of the *latex* document is declared in the second line of the *sample.tex* file. An empty line was inserted for visual clarity after this declaration. To typeset a sentence in italic, we have enclosed it in curly brackets (`{}`) and appended it to the italicizing tag `\textit` in the fourth line. If we wanted to typeset the same sentence in boldface, we would have used the `\textbf` tag. For typewriter face, we would have used the `\texttt` tag. Finally, we mark the end of the document in the last line of the *latex* file.

Equations, graphics, and other elements

A variety of *latex* tags are available for formatting text and composing tables and equations, as explained in books, manuals, and *web* tutorials. For example,

*The capitalization of the *Internet* indicates the word wide web (*web*), as opposed to an arbitrary network.

to typeset the equation

$$E = \frac{a+b}{c+d} \quad , \tag{1.1}$$

including an equation number, we use the following lines:

```
\begin{equation}
  E = \frac{a+b}{c+d}
\end{equation}
```

where *frac* stands for fraction and the rest of the tags have obvious meanings. In fact, this book was typeset in *latex* under the *ubuntu* operating system, and the preceding lines were used verbatim in the source file.

Figures and graphics elements, simple tables, colored tables, long tables, floating schematics, and other components can be easily imported or typeset in a *latex* document. A key idea is the concept of a local environment entered by the `\begin{...}` tag and exited by the complementary `\end{...}` tag, where the three dots represent an appropriate keyword. Examples are the equation and figure environments.

Human readable code

An ordinary person can easily extract and understand the information contained between a *latex* markup tag and its closure. This means that a *latex* document is human readable even when it contains involved equations and other advanced elements, such as tables, schematics, and figures. Graduate students typically learn the basic rules of *latex* typesetting in a few days based on a handed-down template. Experienced *latex* users are able to edit a *latex* document while mentally processing and simultaneously interpreting the typesetting tags and visualizing the final product in their minds. Computer programming experience is not necessary for typesetting a *latex* document. Typing skills and basic deductive ability are the only prerequisites.

Latex tag processing (interpretation)

When and how are the *latex* tags interpreted to produce the final typeset document, such as that appearing on the pages of this book? Once a *latex* source file has been composed, it must be supplied to a *latex* processor. The processor is a program (application) that receives the source *latex* file and produces a binary device-independent file (*dvi*) that can be viewed on a computer terminal using a file reader. In turn, the *dvi* file can be converted into a postscript file (*ps*) or portable document format file (*pdf*) and sent to a printer.* Direct conversion of a *latex* source document into a *pdf* file is possible using a suitable application.

---

*Postscript is a computer language developed for high-quality printing. Postscript interpreters are embedded in postscript printers.

Command-line processing

To process a *latex* file entitled *myfile.tex* and produce the typeset text, we open a terminal (command-line window) and launch the *latex* processor by issuing the statement:

```
$ latex myfile.tex
```

followed by the ENTER keystroke, where the dollar sign ($) is a system prompt. The processor generates a *dvi* file named *myfile.dvi* containing the processed *latex* file. To convert the *dvi* file into a postscript (*ps*) file named *myfile.ps*, we run the *dvips* application by issuing the statement:

```
$ dvips -o myfile.ps myfile.dvi
```

followed by the ENTER keystroke. To convert the postscript file into a portable document format (*pdf*) file named *myfile.pdf*, we may use the *ps2pdf* (*ps* to *pdf*) application by issuing the statement:

```
$ ps2pdf myfile.ps
```

followed by the ENTER keystroke. The application generates a file named *myfile.pdf* in the directory of the source code.

Other methods of producing a *pdf* file from a *latex* file are available. For example, we may use the *pdflatex*, *dvipdf*, or *dvipdfm* applications.

Markup or programming language?

Is *latex* a tagging system (markup language), a programming language, or both? The absence of predefined data structures, the inability to perform arithmetic operations and string manipulations, and the importance of textual content suggest that *latex* is a markup language. As a rule of thumb, a person who is skillful in a *bona fide* computer language can easily learn another *bona fide* computer language. Regrettably or fortunately, depending on the point of view, a person who is proficient in *latex* cannot transition directly to *fortran* or C++. However, typesetting instructions are implemented in a *latex* document and a dedicated processor must be used to implement the *latex* tags and produce the final document. It is fair to say that *latex* is a markup language in some ways, and a lightweight programming language in other ways.

1.1.2 Html

The hypertext markup language (*html*) is used for writing *web* documents that can be processed and displayed in the window of a *web* browser. *Html* tags are keywords enclosed by pointy brackets (<>), also called angle brackets. A perfectly valid complete *html* file may contain the following lines:

```
<html>
    <i>This sentence is set in italic</i>
</html>
```

where <i> and </i> is a pair *html* italicizing tags enclosing data. The forward slash (/), simply called a slash, marks the closure of a previously opened tag.

Interpretation

Html interpreters and viewers are embedded in *web* browsers. The tag <html> at the beginning of an *html* file launches the *html* interpreter, and its closure </html> marks the end of the data to be processed by the *html* interpreter. To interpret an *html* file and view the processed information contained in the file, we may *open* the file by selecting its name in the drop-down *File* menu of a *web* browser. Alternatively, an *html* file can be accessed from an Internet *web* server. To print the processed *html* file displayed in the window of a *web* browser, we may use the browser's printing services typically offered in the *File* drop-down menu.

A wealth of formatting features

Over one-hundred *html* tags are available at the present time, as explained in *html* books and *web* tutorials. Pictures, graphics, tables, music files and videos can be embedded easily in an *html* document using appropriate tags. Processing instructions (PIS) in other programming languages, such as *java* or *javascript*, can be included for the purpose of manipulating or receiving data. For the reasons discussed previously for *latex*, *html* can be regarded both as a tagging system and a lightweight programming language.

Mathml

Equations can be typeset in an *html* document using the mathematical markup language (*mathml*). The following *mathml* code embedded in an *html* document displays equation (1.1) in the window of a *web* browser:

```
<html>
  <math xmlns="http://www.w3.org/1998/Math/MathML">
    <mi>E=</mi>
    <mfrac>
      <mrow>
        <mi>a</mi>
        <mo>+</mo>
        <mi>b</mi>
      </mrow>
      <mrow>
        <mi>c</mi>
        <mo>+</mo>
        <mi>d</mi>
```

```
      </mrow>
    </mfrac>
  </math>
</html>
```

The <mi> tag stands for *mathematical identifier* and the <mo> tag stands for *mathematical object*. The <math> tag in the second line defines an *xml* environment identified by its namespace (*xmlns*), as discussed in Chapter 2. All tags are accompanied by their closure indicated by a forward slash (/) placed at appropriate places to ensure proper nesting.

It is clear that typesetting equations in *mathml* is much more cumbersome than in *latex*. This comparison underscores the superiority of *latex* in technical typesetting and explains why *latex* is a standard choice in technical publishing and academe.

1.1.3 Latex compared to html

Latex and *html* share a number of important features. Both languages are human readable, and both languages convey information (content) and implement presentation (appearance). *Latex* is superior to *html* in that it offers an impressive menu of tagging options, including mathematical equations, tables, and an assortment of special characters and symbols. Almost anything that can be done in *html* can be done in *latex*, but not necessarily *vice versa*. The typesetting quality of *latex* is superior to that of *html*.

Latex is unforgiving

There are penalties to be paid. *Latex* is less forgiving than *html* in that, if a tagging error is made, the processing of the source file will be abandoned by the processor and warnings will be issued and recorded in a log file. In contrast, *html* forgives misprints and ignores minor and sometimes major tagging errors. For example, if a new paragraph is forced to open in an *html* document, the preceding paragraph does not have to be closed. Even if the structure of various tags in an *html* document is arbitrary and nonsensical, information will still be processed and displayed to the best of the ability of the *html* interpreter.

Latex and html compilers are free software

A variety of *web* browsers are freely available for standard operating systems. A *latex* document cannot be interpreted by a *web* browser but requires a dedicated processor. Free *latex* processor and authoring applications are available in a variety of operating systems. The complete *latex* compiler, including an assortment of add-on packages that provide additional functionality, is included in most *linux* distributions, including *ubuntu*. Typesetting a document in *latex* guarantees longevity and compatibility with future media technology.

Latex to html and back

A *latex* document can be converted into an *html* document using a suitable translation program (application), such as *latex2html* or *tth.* Mathematical equations and figures are treated in special ways. A scientist, engineer, or technical typesetter may write a *latex* document, convert it into an *html* document, and post it on the Internet for direct viewing. The ability to convert a *latex* document into an *html* document hinges on the consistent use of corresponding formatting tags. Although it is also possible to convert an *html* document into a *latex* document, in most cases the effort is a mere academic exercise.

Exercises

1.1.1 *Latex, html, and mathml*

(*a*) Write a *latex* file that prints your favorite color in boldface. (*b*) Repeat for an *html* file. (*c*) Write an *html* file that prints the equation $E = mc^2$ using *mathml.*

1.1.2 *Latex, html, and mathml tags*

(*a*) Prepare a list of sixteen *latex* tags of your choice. (*b*) Repeat for *html* tags. (*c*) Repeat for *mathml* tags.

1.2 Formatting with xml

In both *latex* and *html*, the menu of available tags is determined by the authors of the respective markup language parser and interpreter.* In the case of *latex*, the interpreter is the *latex* processor. *Html* interpreters are embedded in *web* browsers.

In writing a *latex* or *html* document, we must strictly adhere to standard structures and conventions decided by others. For example, the tag `\textlatin` is meaningless and will produce errors in *latex*, and the tag `<puppy>` is meaningless and will be ignored in *html.* If tags for typesetting equations were not available in *html*, the language would be of limited use to quantitative scientists and engineers.

Computer programmers, document typesetters, and data-entry operators welcome the opportunity of using formatting tags that best suit their practical or creative needs in different applications. The generalized framework implemented by the extensible markup language (*xml*) allows us to employ tag names and structures that best describe the components of a document of interest and the individual pieces of data contained in a database. Of equal importance,

*In computer science, *parsing* describes the process of breaking down a chain of words into elementary pieces (tokens) and applying a set of rules to recognize instructions and extract attributes and parameters.

xml facilitates the unique identification of similar pieces of data so that targeted information can be readily extracted from a database, as the need arises.

To illustrate the concept of data identification, let us assume that the title of a book chapter is set in bold face in an *html* document using the `<b>` tag and its closure `</b>`. Because the `<b>` tag could also be used to emphasize the ISBN number, we are unable to identity the book title by searching through the document for the bold face tag in the absence of further identifying information. In an *xml* document, tags that unambiguously and uniquely define the beginning, end, and title of each chapter are employed.

Xml keywords are arbitrary

Xml allows us to use any desired, but sensible and consistent, tagging system that best suits our needs. Arbitrary and unrelated names regarded as user-defined keywords can be assigned to the individual tags, and optional qualifiers known as attributes can be employed. A typical *xml* document contains sequences of nested tags describing and evaluating a multitude of objects and structures without any constraints, apart from those imposed by basic *xml* grammar, as discussed in Chapter 2.

These features render *xml* a meta-language; a better term would be a flexible language; an even better term would be a flexible tagging system. However, in all honestly, *xml* is not a language, but an *adaptable data formatting system* that can be interpreted by a person or processed in unspecified ways by a machine. *Xml* appears as a language only when compared to a spoken world language that evolves in response to new concepts, terms, and emerging communication needs.

Flowers

To record and describe a flower in an *xml* document, we may introduce the flower and list its properties:

```
<flower>
  <kind>rose</kind>
  <color>red</color>
  <smell>captivating</smell>
</flower>
```

Like *html* tags, *xml* tags are enclosed by pointy brackets (`<>`), also called angle brackets. As in *html*, a closing tag arises by prepending to the name of the corresponding opening tag a forward slash (/), simply called a slash.

In our example, the flower is an object, identified as an *xml* element, whose properties are recorded in a nested sequence of tags, identified as children elements with suitable names. The opening tag `<flower>` is accompanied by the corresponding closing tag `</flower>` to indicate that the flower description has ended. All other tags inside the parental flower tag open and close in similar ways. Three pieces of data are provided and simultaneously described in this document: rose, red, and captivating. Other flowers can be added to describe a bouquet in a living room or flower shop.

In an alternative representation, the flower of interest can be described in terms of attributes in a single line as:

```
<flower kind="rose" color="red" smell="captivating" />
```

In this formulation, `kind`, `color`, and `smell` are attributes evaluated by the contents of the double quotes constituting the data. For reasons of scalability and ease of retrieval, the expanded representation in terms of nested tags is preferred over the attribute representation, as discussed in Chapter 3.

Data organization

The flower example illustrates two important features of the *xml* formatting system: a high level of organization, and unique data identification. A *latex* document also exhibits a high level of organization. However, *latex* and *html* are primarily concerned with data presentation in printed or electronic form. In these restricted tagging systems, information retrieval is only an afterthought.

Data formatting with xml does not require programming experience

Programming language skills are not necessary for composing and editing an *xml* document containing data. The document can be written by a person in isolation following basic *xml* grammar, as discussed in Chapter 2, with no reference whatsoever to conventions imposed by others. Common sense, editorial consistency, typing skills, and a general plan on how the data will be organized are the only prerequisites.

Text (ascii) files

To generate an *xml* document, we write a *text* file, also called an *ascii* file, using a word editor of our choice, such as *nano*, *pico*, *emacs*, *vi*, or *notepad*. The names of the formatting tags can be words of the English language or any other spoken or fictitious language. Advanced word processors, such as *LibreOffice Writer*, can be used, but the file must be saved as an unformatted *text* or *ascii* document.

Ascii is an acronym of the American standard code for information interchange. A *text* or *ascii* file contains a sequence of integers, each recorded in 7 binary digits (*bits*) in terms of its binary representation. The integers represent characters, including letters, numbers, and other symbols, encoded according to the *ascii* convention discussed in Appendix A. Characters outside the *ascii* range may also be used according to generalized character encoding systems, as discussed in Section 2.4.

Text or *ascii* files contain long binary strings consisting of 0 and 1 digits describing integers that can be decoded by a person or application with reference to the *ascii* map. An example is the string

```
0100110 0111000 1101100 0101110 ···
```

consisting of a chain of seven bits. In contrast, a *binary file* contains binary strings encoding machine instructions, data formatting, and other information pertinent to a specific application, such as a spreadsheet. A binary file can be opened and processed only by its intended application.

Fasolia and keftedakia

If we want to record a sentence in italic in an *xml* document describing a delicious meal (beans and meatballs), we may write:

```
<set_in_italic> Fasolia and keftedakia </set_in_italic>
```

This line can be part of an *xml* file describing a dinner menu. The underscore (_) is used routinely to connect words into a sentence that enjoys visual continuity as an uninterrupted character string. Who will interpret the italicizing tags is of no interest to the reclusive author of this *xml* document. In contrast, mandatory italicizing tags must be employed in *latex* and *html*documents, as discussed in Section 1.1.

Nuts

Following is a complete *xml* file containing a prologue (first line) and a list of nuts:

```
<?xml version="1.0"?>
<pantry>

  <nut>peanuts</nut>
  <nut>macadamia</nut>
  <nut>hazelnuts</nut>

</pantry>
```

The prologue is an *xml* document declaration inserted in most *xml* files, as discussed in Section 2.4. It is clear that the data contained in this file represent three nuts found in a pantry. The tag `<pantry>` defines the *root element* of this *xml* document, enclosing all other children elements implemented by opened and closed tags. Note that all tags close at expected places. Additional nuts can be easily removed, added, or replenished. Further information on how many nuts of each type are available could have been included as properties represented by children elements or element attributes. The question of what to do with these nuts is outstanding.

Equations

A complete *xml* document describing equation (1.1) may contain the following lines:

```
<?xml version="1.0"?>
<equation>

  <left_hand_side>
    E
  </left_hand_side>

  <right_hand_side>
    <fraction>
      <numerator>
        a+b
      </numerator>
      <denominator>
        c+d
      </denominator>
    </fraction>
  </right_hand_side>

</equation>
```

The tag `<equation>` defines the root element of this *xml* document, enclosing all other children elements. This *xml* document can be converted into an equivalent *latex* or *mathml* document manually or with the help of a suitable computer program (application).

An xml document presents and describes data

An extremely important feature of *xml* is that data, including numbers, objects, items, instructions, and statements of a computer programming language, are not only *presented*, but also *described* in a chosen language of the world, such as English or Portuguese. In fact, the data can be written in one language, and the tags can be written in another language. This duality is intimately related to the desirable property of plurality in a human readable database or code.

Data trees

A conceptual tree can be built expressing precisely and unambiguously a hierarchy of information in an *xml* document. One example is a tree describing all parts of a car arranged in branches identified by tags named `engine`, `wheels`, `cabin`, and other components. Another example is a tree describing the elements employed in a finite-element code. A third example is a tree describing orthogonal polynomials, distinguished by their domain of definition and weighing function. A fourth, less apparent but more intriguing example, is a tree of statements of a computer programming language implementing a numerical algorithm.

Data parsing

We have mentioned that data, instructions, and other information encapsulated in an *xml* document can be read and understood by a person or else inspected (parsed) and processed by a computer program (application) written in a language of our choice, such as *fortran*, C, C++, *java*, *perl* or *python*. *Xml* parsers are available as modules of advanced computer languages, including C++, *java*, *javascript*, *php*, *perl*, and *python*. Relevant procedures for selected languages will be illustrated in Chapter 5.

An xml document must be well-formed

Although the names of the tags employed in an *xml* document can be arbitrary, the document itself must be well-formed.

One requirement is that an opening tag defining an element, such as `<nut>`, be accompanying by the corresponding closing tag `</nut>` to indicate the end of a nut. The closing of a tag is mandatory even when it appears unnecessary, as in the case of a tag forcing a line break in a word document.

For an *xml* element to be well-formed, tags must be properly nested, as discussed in Chapter 2. This means that two nuts may not overlap, that is, one nut may not cross over another nut.

However, these restrictions are mild and reasonable constraints imposed to ensure successful data parsing, prevent confusion, and avoid misinterpretation. Identifying grammatical errors in an *xml* document is straightforward. Only when an *xml* document is exceedingly long is the help of a computer processor (*xml* debugger) necessary. In contrast, identifying bugs in a computer code can be tedious and time consuming. In some cases, it may take a few hours to write a computer code and then weeks to remove fatal or benign errors.

Visualization in a web browser

We can open an *xml* file with an *xml* compliant *web* browser through the *Open File* option of the *File* drop-file menu. In the absence of processing instructions (PI) embedded in the *xml* file, as discussed in Chapter 2, the *xml* data tree will be visualized with a ± mark on the left margin. Clicking on this mark with the mouse will reveal or hide the branches of the *xml* tree. A warning will be issued if the *xml* file is not well-formed.

Xhtml

Roughly speaking, an *xhtml* document is a well-formed *html* document where all tags are written in lower case. Only established *html* tags can be used in an *xhtml* document, that is, improvised tags cannot be employed. An *xhtml* document is also an *html* document, but an *html* document is not necessarily an *xhtml* document. An *html* document can be certified as an *xhtml* document by a suitable program called an *xhtml* validator.

Is a latex or xhtml document an xml document?

Although *latex* and *xhtml* documents must be well-formed, the tagging system lacks the necessary flexibility and extensibility that is the hallmark of the *xml* layout. Most important, arbitrary tags and tag attributes cannot be added at will.

However, any *xml* document that is bound to an agreed convention also suffers from inflexibility and inextensibility. This observation raises a concern as to whether *xml* lives up to its advertised quality as a genuine meta-language in practical applications tied to industry standards. This well-founded skepticism will be revisited throughout this book.

Xml authoring tools

Xml editors are word editors (applications) with a graphical user interface (GUI) that highlights with color the tagging tree of an *xml* document. The objective is to help ensure that the document is well-formed by approving or dismissing the tagging structure employed. These authoring tools are helpful but not necessary for composing an *xml* document.

Computer language implementations

We have discussed data formatting and physical or abstract object description in the *xml* framework. In fact, *xml* can be used to implement computer language instructions. Tags with attributes in an *xml* compliant programming language play the role of logical and other constructs. Examples are the *for* loop in C++ and the *Do* loops in *fortran*. Details will be given in Section 1.3.

Constraints

Constraints on the tagging system of an *xml* document arise only when the *xml* data are written to be sent to another person or application. The goal of these constraints is to ensure that sender and receiver agree on the amount and type of information contained in an *xml* document of interest to both. Not surprisingly, *xml* formatting becomes relevant to scientific computer programming only with regard to instruction syntax and formatting of input/output (I/O).

Exercises

1.2.1 *Tools in a shop*

Write an *xml* document that lists the tools in a carpentry shop along with other pertinent tool information.

1.2.2 *Books in a shop*

Write an *xml* document that lists the books in a bookshop along with titles, authors, and year of publication.

1.2.3 *Orthogonal polynomials*

Write an *xml* document that describes three families of orthogonal polynomials of your choice using information and tags of your choice.

1.3 Usage and usefulness of xml files

What can we do with a well-formed *xml* document containing useful data, statements, or computer language instructions implemented by nested pairs of opened and closed tags and optional attributes? A few general but related families of applications are possible.

In reviewing these applications, it is helpful to make a distinction between *xml* data files and *xml* program files. An *xml* data file contains data and possibly processing instructions but no code. A human or machine processor is needed to manipulate the data and display the outcome in some desired way. In contrast, an *xml* program file contains instructions of a suitable computer language, such as the *xsl* language discussed in Chapter 3. Statements of any computer language can be recast in the *xml* format using appropriate tags and attributes.

It should be mentioned at the outset that the usage and usefulness of *xml* files can be understood fully only after *xml* data manipulation has been demonstrated and specific applications have been discussed. Realizing the purpose and utility of *xml* requires patience and a certain degree of hindsight.

1.3.1 Data formatting

The vast majority of *xml* applications are concerned with data formatting. In these applications, the structure of the tagging system in an *xml* document (data tree) is designed carefully to hold desired pieces of information, and at the same time avoid redundancy and repetition while anticipating future needs. A reputable *xml* designer will draw a bicycle that could be extended into an automobile, if the need arises, unrestricted by physical exclusion: wheels and engine will not overlap.

Conversion of an xml data file into another formatted data file

An *xml* data file can be transformed into another formatted data file. For example, an *xml* file can be converted into an *html* file whose content can be viewed on a *web* browser and then printed on paper, as discussed in Section 1.5. In the process of conversion, data can be manipulated and calculations can be performed to produce new data or suppress unwanted data. To transfer information from one application into another, data can be extracted from a source *xml* file, reformatted, and recorded into a new file or embedded into a new application under a different tagging system. *Xml* data documents written with ease-of-conversion in mind are sometimes called document-centric.

Data transmission

An *xml* data file created manually by a person or automatically by an application (program) can be sent to another person or application to be used as input. For example, data generated by a spreadsheet can be stored in an *xml* file under agreed conventions, and then imported into another application using different conventions. In fact, an *xml* document can serve as an interface between applications with different native formats: *pdf* may be converted into *xml* and then imported into a spreadsheet, and *vice versa.*

In scientific computing, data can be extracted from an *xml* file and automatically accommodated into variables or arranged into vectors and matrices (arrays) suitable for numerical computation. For example, an *xml* file may describe the geometrical properties of the elements of a boundary-element simulation along with a computed solution.

Data retrieval

The information contained in a small or large *xml* data file is a database. Data of interest contained in this database can be extracted using a general or special-purpose computer language code. For example, the year of publication can be retrieved from a document containing a list of books or research articles. *Xml* data written with ease-of-retrieval in mind are sometimes called data-centric.

Standardization

Numerous *xml* formatting systems have been proposed for data specific to particular disciplines: from music, to transportation, to computer graphics, to science and engineering applications. A comprehensive list of established tagging systems is available on the Internet.* New tagging systems consistent with *xml* conventions are frequently introduced. It is generally accepted that *xml* is the default framework for data formatting and storage applications.

*http://en.wikipedia.org/wiki/List_of_XML_markup_languages

1.3.2 Computer code formatting

Computer languages whose instructions are implemented in the *xml* format have been developed. Corresponding language compilers or interpreters, generically called a language processors, are necessary. In the absence of a corresponding language processor, a computer code written in *xml* is useful only as a prototype. Specific examples of *xml* computer language instructions will be discussed in Section 2.12 after the basic rules of *xml* grammar have been outlined.

Chula_vista

As a preview, we consider the following instructions in a fictitious computer language called *chula_vista*:

```
Do_this_for i=1:1:10
  display_on_the_screen i*i
End_of_Do_this_for
```

where the asterisk indicates multiplication. These lines print on the screen the square of all integers from 1 to 10. The same instructions could have be encoded in the *xml* format in terms of judiciously selected tags, as follows:

```
<chula_vista:Do_this_for variable="i" low="1" high="10" increment="1">
  <chula_vista:display_on_the_screen select="square(i)"/>
</chula_vista:Do_this_for>
```

Note that the name of the computer language employed, *chula_vista*, is specified for clarity and completeness in each *xml* tag along with necessary attributes. In standard *xml* nomenclature, the keyword *chula_vista* is a namespace. A self-closing tag is employed in the second line where the function *square* is called with a single argument to evaluate an attribute.

It is striking that the native *chula_vista* code is cleaner than its equivalent *xml* implementation. This observation confirms our suspicion that the *xml* formatting protocol is not without shortcomings.

Beware of exaggerations

The usefulness of *xml* in coding computer language instructions will be questioned by scientific programmers. The main reason is that computer code in any mid- or upper-level *bone fide* language, such as *fortran* or C++, is human readable by design. Low-level assembly code is human readable to a lesser extent, whereas machine code is incomprehensible to the casual computer programmer. It could be argued that an *xml* compliant code may serve as a generic blueprint that can be translated into any other computer language code. However, the same is true of *fortran*, C++, or any other upper-, mid- or low-level programming language equipped with appropriate data structures and language constructs.

Xml is appropriate for multilingual code

The *xml* implementation of a programming language is useful in cases where a computer code contains instructions in different programming languages, or employs functions implemented in different linked libraries. Consider the following *xml* implementation of the *chula_vista* code:

```
<chula_vista:Do_this_for variable="i" low="1" high="10" increment="1">
  <chula_vista:display_on_the_screen select="smolikas:gamma(i)"/>
</chula_vista:Do_this_for>
```

In this case, the `smolikas:gamma()` function belonging to the *smolikas* namespace is employed to compute the Gamma function.

In Chapter 3, we will see that multiple languages are used extensively in the *web* processing of *xml* and *html* files. Scientific programmers are used to writing code in one chosen language with occasional cross-over to other languages by way of language wrappers.

Exercises

1.3.1 *Multiple use of data*

Discuss possible ways of reusing text recorded in a question-and-answer (Q&A) session following a lecture recorded in an *xml* file.

1.3.2 *Eternity*

Write a sensible code of your choice in a fictitious language called *eternity*, implemented according to *xml* conventions.

1.4 Constraints on structure and form

To ensure that a sender and a receiver of an *xml* file interpret the data contained in the *xml* file in the same way, the meaning and structure of the formatting tags and data types employed should be defined in anticipation of present and future needs. In addition, other sensible or desirable constraints should be imposed to comply with industry standards. For example, we may require that each chapter of a book has at least one section, and the year of publication of a cited article is entered in the four-digit format. Ten digits would be required if the *xml* document is expected to survive after our sun implodes.

Data type definitions and schema

To achieve these goals, we introduce a document type definition (*dtd*) or an *xml* schema definition (*xsd*). In practice, we write instructions that define the meaning and prescribe the permissible structure of tags to be employed, and also introduce constraints on data types, as discussed in Section 2.10. These instructions are either embedded into an *xml* data document or placed in an accompanying document. The recommended and most powerful method of implementing an *xml* data type definition is the *xml* schema. Initiatives are under way to develop schemata in various branches of commerce and publishing for the purpose of standardization.

Cml and xdmf

As an example, the *xml* schema definition (*xsd*) of the chemical markup language (*cml*) specifies the following typical structure:

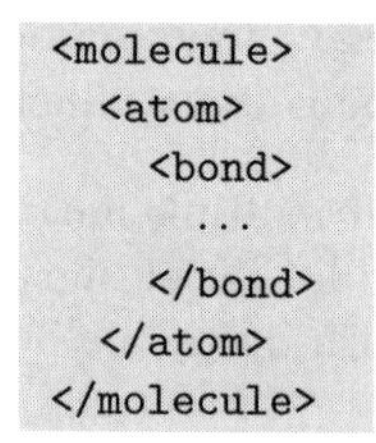

```
<molecule>
  <atom>
    <bond>
      ...
    </bond>
  </atom>
</molecule>
```

where the three dots indicate additional data. Similar structures and conventions must be followed in documents that comply with the extensible data model and format *xml* schema (*xdmf*) designed for high-performance computing.

Validation

Once a *dtd* or *xsd* has been selected and implemented, a complying *xml* document must employ only mandatory or optional tags and structures defined in the *dtd* or *xsd*. An *xml* data file can be validated against a *dtd* or *xsd* using a standalone validation program or a validation program included in an *xml* parser. It is important to remember that an *xml* file that fails to be validated may still be well-formed.

Loss of freedom

By accepting a *dtd* or *xsd*, we abandon our freedom to compose and format a document using tag names and data structures of our choice. To be well-formed is no longer sufficient but only necessary. Our status regresses to that of a *latex* or *html* user who must use predetermined tags implemented in the *latex* processor or *html* interpreter. Sadly, we may no longer write an *xml* document in isolation, but must pay attention to an established set of rules. Although programming experience is still not necessary, the use of a reference manual is mandatory.

Snake oil?

We have reached a crossroads and it appears that we have made a full circle. In light of the potential loss of freedom incurred by adopting a *dtd* or *xsd*, it is fair to question whether *xml* lives up to its reputation as a panacea. The truth is that adopting a standard *dtd* or *xsd* makes sense when the formatting protocol is broadly accepted and well documented, or when the source code of the *dtd* or *xsd* is accessible and freely available for modification. Alternatively, the lone researcher, scientific programmer, research group, or computer enthusiast may build their own private set of rules without any constraints. *Xml* formatting can become a personal way of recording thoughts and keeping notes.

1.4.1 DocBook schema

It is instructive to discuss an example of a document type definition used in a popular application. The *DocBook* schema was designed for writing books and other documents following *xml* conventions.* *LibreOffice*† is a free authoring application available on a variety of platforms, incorporating the *DocBook* schema. The suite includes a *what you see is what you get* (WYSIWYG) word processor named *writer* that is compliant with the *DocBook* schema.

As a digression, we note that WYSIWYG editors have been criticized for dividing an author's attention into substance and form. The main argument is that an author should be encouraged to write complete, thoughtful, and well-structured sentences undistracted by formatting and spelling considerations, and then format or embellish the presentation. This is precisely what *latex* seeks to accomplish.

The Great Gatsby

As an experiment, we launch (open) the *LibreOffice* word processor and use the available menu to write the following lines from F. Scott Fitzgerald's *The Great Gatsby*:

Chapter 1

In my younger and more vulnerable years my father gave me some advice that I've been turning over in my mind ever since.

"Whenever you feel like criticizing any one," he told me, "just remember that all the people in this world haven't had the advantages that you've had."

Every word, every sentence, every punctuation mark in this passage is outstanding. Nothing can be improved in style, meaning, presentation, or intent. F. Scott Fitzgerald was a brilliant writer. Novelists, journalists, and technical writers will benefit a great deal from reading his books.

Next, we select *Save As* from the *file* drop-down menu of the application, choose *DocBook* as Filetype, and save the file under the name *greatgatsby.xml*.

---

*http://www.docbook.org/whatis

†http://www.libreoffice.org

The content of this file is:

```
<?xml version="1.0" encoding="UTF-8"?>
<!DOCTYPE article PUBLIC "-//OASIS//DTD DocBook XML V4.1.2//EN"
"http://www.oasis-open.org/docbook/xml/4.1.2/docbookx.dtd">
<article lang="">
  <para>Chapter 1</para>
  <para/>
  <para>In my younger and more vulnerable years my father
  gave me some advice that</para>
  <para>I've been turning over in my mind ever since.</para>
  <para>  ''Whenever you feel like criticizing any one,''
   he told me, ''just remember</para>
   <para>that all the people in this world havent
  had the advantages that youve had.'' </para>
</article>
```

The name of the *xml* root element, `article`, is qualified by an empty (default) language attribute named `lang`. The meaning of other tags implementing children elements can be deduced by mere inspection and sensible interpretation. For example, `para` is an abbreviation of *paragraph*. The second line, continuing to the third line, reading:

```
<!DOCTYPE ··· docbookx.dtd">
```

invokes a document type definition (*dtd*), as discussed in Section 2.10.

The *LibreOffice* native file itself, named *greatgatsby.odt*, is a machine readable binary file that can be interpreted only by the word processor. In contrast, *greatgatsby.xml* is a *text* (*ascii*) file.

1.4.2 LibreOffice Math

The *LibreOffice* suite includes an equation editor application named *math*. As an experiment, we open the application and type in the lower partition of the graphical user interface (GUI) the following text:

```
{a} over {b}
```

representing the fraction a/b. Next, we save the text in a file with a chosen name. The content of this file turns out to be:

```
<?xml version="1.0" encoding="UTF-8"?>
<math xmlns="http://www.w3.org/1998/Math/MathML">
  <semantics>
    <mrow>
    <mfrac>
      <mrow>
        <mi>a</mi>
      </mrow>
```

```
        <mrow>
          <mi>b</mi>
        </mrow>
      </mfrac>
      </mrow>
    <annotation encoding="StarMath 5.0">a over b </annotation>
    </semantics>
  </math>
```

This is an *xml* file with a root element named `math` associated with a namespace specified in the *xmlns* attribute. We observe several nested tags with apparent or nearly obvious meanings. In fact, this file can be opened with a *web* browser to display the fraction.

This example illustrates the portability of *xml* data across different applications running on arbitrary hardware platforms. In the absence of proprietary encoding that obscures the meaning of information, an *xml* document can be created by one application and then imported unchanged or slightly modified into another application. This property is sometimes described as *information reuse* or *multiple data use.*

Exercise

1.4.1 *An experiment*

Repeat the *DocBook* experiment discussed in the text with a document of your choice.

1.5 Xml data processing

We have mentioned on several occasions that data contained in an *xml* file can be read, understood, and interpreted by a person or else processed by an application written in a suitable computer language. Specific examples are given in this section.

1.5.1 Human processing

A busy financier fills out a lunch order on an *xml* order form and faxes it to a restaurant. The restaurant owner visually parses the order and delivers instructions to the cook who follows the instructions and prepares the food. The order is also handed to the cashier who assesses charges and taxes, prepares a bill, and faxes a bill to the financier. In this case, the owner of the restaurant is the parser, the cook and the cashier are two different processors, and the financier has a data-entry job.

1.5.2 *Machine processing with xsl*

The computer processing of data contained in an *xml* document is best explained in the context of the extensible stylesheet language (*xsl*) discussed in detail in Chapters 3 and 4.

Xsl is a computer language comparable to *fortran* or C. An *xsl* processor, like any other *xml* processor, parses the data contained in a well-formed and possibly validated *xml* document (input), performs calculations and manipulations according to instructions given in a companion *xsl* program file containing code (application), and records or displays the outcome (output). The output can be another *xml* file or an *html* file that can be processed and displayed in a *web* browser. Available *xsl* processors are reviewed in Section 3.1.

We will see that, strange though it may seem, an *xsl* code is written according to *xml* conventions, that is, *xsl* is an *xml* implementation. However, the *xml* compliance of *xsl* is not an essential feature of the *xml*/*xsl* framework.

Sweet and sour

Xsl codes contained in two different *xsl* files may assign different meanings to a tag named, for example, *flavor*, in an *xml* file. One *xsl* file may interpret the tag as sweet, while another *xsl* file may interpret the tag as sour. Thus, depending on the instructions given in the companion *xsl* file, an *xml* data file with the same words may taste differently after processing,

Debt and donation

Assume that an *xml* file contains the names and details of university graduates. These data can be used to send the graduates two letters: a fundraising letter asking for donations, and another letter asking them to pay owed tuition and fees. The primary *xml* file (input) reads:

```
<graduate>
  <name>Eliana Smith</name>
  <major>entomology</major>
  <graduation>2003</graduation>
  <debt>87451.20</debt>
<graduate>
```

After processing, the output file relevant to the fundraising letter may read:

```
<donor>
  <name>Eliana Smith</name>
  <major>entomology</major>
  <years_since_graduation>8</years_since_graduation>
  <recommended_donation>20.0</recommended_donation>
<donor>
```

The output file relevant to the debt-collection letter may read:

```
<bill tone="harsh">
  <name>Eliana Smith</name>
  <you_owe_us_with_interest>999817.99</you_owe_us_with_interest>
  <pay_by>yesterday</pay_by>
<bill>
```

Some calculations were performed in generating the output files. This example illustrates that an *xml* document may serve as an information database in a multitude of applications.

Xsl constituents

The *xsl* processor encapsulates three interacting libraries with complementary tasks:

- *Xslt* (transformations) is a language for transforming an *xml* document (input) into another *xml* document (output), including an *html* document or any other text (*ascii*) document.
- *Xpath* is a language for navigating inside an *xml* document by returning references to *xml* element nodes.
- *Xsl-fo* is a language for formatting an *xml* document.

Xslt allows us to rearrange, suppress, modify, and add to the information contained in the *xml* data file, as discussed in Chapters 3 and 4. *Xslt* uses *xpath* to match and select coherent data blocks. When a match is found, *xslt* applies the requested transformations. An *xsl* procedure or function is sometimes delineated as an *xslt* or *xpath* element or function to accurately describe its implementation.

Since the internal organization of the *xsl* processor is of marginal interest in scientific computing, we will generally refer to *xsl* in place of *xslt* or *xpath* when a distinction is not necessary.

Xsl computing

The *xsl* processing of an *xml* file follows the paradigm of scientific computing in that data and programs are separated into different files. However, significant differences in programming structures and available facilities render the two frameworks sharply distinct. In particular, a number of programming structures shared by common scientific languages, such as *fortran* or C, do not have counterparts in the *xsl*, and *vice versa*. The reason can be traced to the paramount importance of the data in the *xml*/*xsl* framework. We will see that the mere presence of an *xml* tag is sufficient to drive the execution (launching) of an *xsl* code.

Bare bones

Because of severe limitations with regard to numerical procedures and functions, the *xml/xsl* framework is not suitable for advanced scientific computing. Arguments to the contrary lack convincing counter-examples. Nonetheless, the unavailability of extensive resources renders the *xml/xsl* framework attractive for developing programming skills, as discussed in Chapter 4. Metaphorically speaking, the *xml/xsl* framework provides us with a screwdriver and some screws and expects us to build a John Deere tractor.

Execution begins with the data

It is worth remarking that the execution of an *xsl* code begins with the data rather than with the code. Specifically, the name of the file hosting the *xsl* code is defined in a processing instruction in the *xml* data file. This feature is consistent with the notion that the data are more valuable than the program that manipulates the data. In scientific computing, a reliable program that computes the eigenvalues of an arbitrary matrix is extremely valuable. In the *xml* framework, the words of *The Catcher in the Rye* are precious, irrespective of how they appear printed on paper or displayed on a screen.

Xml is a formatting language, xsl is a computer language

Most *xml* texts and *web* tutorials discuss exclusively the *xml/xsl* framework, and this may create the misconception that *xml* is intimately connected to *xsl* or *vice versa.* This is certainly not true in the context of scientific computing where the *xsl* language is hardly known. In its pure and intended form, *xml* is a general and unrestricted data or statement formatting protocol.

Xml to html

An *xsl* code can be written that converts an *xml* file into an *html* file that can be stored in a file or processed and viewed in a *web* browser. In the simplest and most direct method, the *xml* file is opened by a *web* browser through a drop-down menu. A processing instruction (PI) near the beginning of the *xml* file indicates the name of an accompanying *xsl* file that will be used to process the *xml* data. If this file is not found, an error message is issued.

If the file is found, the *xsl* parser and processor embedded in the *web* browser process the data and display the outcome in the browser's window. To print the processed *html* file, we use the browser's printing services. Since all modern browsers can handle basic *xsl* code, *xml* processing is independent of the computer platform employed. This independence is the cornerstone of the *xml* framework within and beyond the *xsl* framework.

Xml was motivated to a large extent by the desire to generalize the restrictive *html* framework in *web* programming. This motivation is of marginal interest in scientific computing where the ability to generate *html* code from *xml* code is hardly compelling. The main attraction of *xml* in computational science relates to its ability to describe, organize, and identify data in the input or output.

Xml processors are strict

The world wide *web* consortium (W3C) specifies that, if an error is found in an *xml* document, the execution of a program or application processing the data, such as an *xsl* code, should terminate. If the execution does not terminate, the code is not W3C compliant. Thus, like *latex* and computer language compilers, but unlike *html* interpreters, *xml* processors are required to be strict.

Consequently, although programming experience is not necessary for composing and editing an *xml* document, syntactic and structural errors are not allowed.

Exercise

1.5.1 *Point particle trajectory*

A data file contains in four columns the three Cartesian coordinates of a point particle in space at a sequence of times. How would this file appear in the *xml* format?

1.6 Relevance of xml in scientific computing

Our main goal in this book is to discuss the relevance of *xml* in scientific computing, which can be contrasted with *web* and professional applications computing. The *xml* framework is consistent with the standard protocol of scientific computing involving input, processing, and output.

When we write a scientific code in a compiled computer language, such as *fortran*, C, or C++, we follow three basic steps:

1. We write a program containing the language instructions.
2. We compile the program and link all object files and necessary libraries into an executable file, which is a binary file containing standalone machine language instructions.
3. We run the executable.

The input data can be contained in the program, entered manually through the input devices (keyboard and mouse) during execution, or read from input data files named in the program. Interpreted code is handled in similar ways.

Value of the executable

In scientific computing, a great deal of effort is expended toward generating efficient code compiled into an executable. In parametric investigations, the same program is run with different input to generate reliable output that can be analyzed by post-processing, visualization, or animation. The central goal of the discipline of high-performance scientific computing is to develop efficient mathematical, numerical, and memory management algorithms that reduce the demand on hardware and central processing unit (CPU) time.*

Separating program from data

In scientific computing, we routinely separate the computer program (code) from the input data, placing them in different files. The program is written in a main file accompanied by function or subroutine files recognized by appropriate suffixes, such as .f, .c, .cc, and .sce, for *fortran*, C, C++, and *scilab*, respectively. The input data reside in other files recognized by standard or arbitrary suffixes, such as .dat, .conf, or .inp. The output is printed in data files named in the code and produced during the execution.

If we change the input data or parameters inside a code, we must be recompile the code to generate a new executable. This major inconvenience explains why it is highly desirable for the data to be separated from code. In the basic *html* implementation, formatting instructions and data reside in the same file. This monolithic structure considerably complicates data extraction, manipulation, reformatting, and portability across hardware platforms. One advantage is that only one computer file needs to be edited, processed, or communicated.

Data formatting

Assume that a data file contains in two columns the real and imaginary parts of the eigenvalues of a 2×2 matrix, as follows:

```
 0.134 -0.234
-0.878  0.238
```

An *xml* file might represent these data as:

```
<eigenvalue>
  <real>0.134</real>
  <imaginary>-0.234</imaginary>
</eigenvalue>

<eigenvalue>
  <real>-0.878</real>
```

*Pozrikidis, C. (2008) *Numerical Computation in Science and Engineering.* Second Edition, Oxford University Press.

```
    <imaginary>0.238</imaginary>
</eigenvalue>
```

The *xml* file describes the eigenvalues unambiguously, circumventing the need for legends, explanations, and conventions. This example underlines the notion that *xml* presents *and* describes data. We say that *xml* encapsulates content and form.

Verbosity is a concern

There is an elephant in the room: the use of repetitive tags in an *xml* document is a practical concern with regard to document size. In our example, the text file containing the matrix of eigenvalues is much smaller than the *xml* file where repetitive tags are employed. Inflated storage can be tolerated in some, but not all, scientific applications. To address this concern, a binary characterization of *xml* data (*xbc*) has been proposed.

Let's be honest

In fact, *xml* is not the only method of presenting and describing data. Other methods based on high-level computer languages are available, as discussed in the remainder of this section. An important requirement is computer programming experience. An important concern is that the principle of separating code from data is likely to be violated. It is not surprising that the uninvested scientific programmer will linger between spending time and effort in developing *xml* compliant schemes or staying in the mainstream.

1.6.1 Matrices

In scientific computing, we routinely deal with matrices defined as square, rectangular, or slender arrays of numbers. An example is the 3×4 matrix

$$\mathbf{A} = \begin{bmatrix} 1 & 2 & 3 & 4 \\ 1 & 4 & 9 & 16 \\ 1 & 8 & 27 & 64 \end{bmatrix}, \tag{1.2}$$

containing 3 rows and 4 columns. In *fortran*, the elements of this matrix are denoted as

```
A(i,j)
```

where `i = ` $1, 2, 3$ and `j = ` $1, 2, 3, 4$. For example, $\mathrm{A}(3, 2) = 8$. In C++, the elements of this matrix are denoted as

```
A[i][j]
```

For example, $\mathrm{A}[3][2] = 8$.

In *fortran*, the indices of matrix elements can have any positive, zero, or negative values. This flexibility considerably simplifies the implementation of algorithms in science and engineering applications. In C and C++, the indices of matrix elements can have positive or zero, but not negative, values. If we reserve a 10×9 matrix $\mathbf{A}$ in a C++ code, we will be allowed to use the matrix elements `A[i][j]`, where `i` $= 0, 1, \ldots, 9$ and `j` $= 0, 1, \ldots, 8$. The zero lower limit ensures the efficient use of all available memory space associated with the binary representation. In *Matlab*, the indices of a matrix element must be positive integers.

Limitations

Two important limitations of the matrix (array) representation of the data shown in (1.2) can be identified:

1. The meaning of the twelve numbers encapsulated in the matrix is not revealed in the matrix itself, but must be separately specified.
2. The elements of the matrix must be either all integers or all real numbers, that is, they must all be of the same data type.

The second difficulty can be resolved by transforming the matrix into an object described by properties and attributes with different data types.

1.6.2 Objects

Consider the following generalized matrix containing numbers and text, regarded as an inhomogeneous object:

$$\mathbf{B} = \begin{bmatrix} 1 & \text{triangle} & 3 & 0.0 & 0.0 & 0.0 & 1.0 & 1.0 & 0.0 & & \\ 2 & \text{square} & 4 & 0.0 & 0.0 & 1.0 & 0.0 & 1.0 & 1.0 & 0.0 & 1.0 \end{bmatrix}. \qquad (1.3)$$

Although it appears safe to assume that this matrix holds information on polygons whose vertex coordinates are provided as consecutive xy pairs, our intuition could be wrong. The first column appears to host a counter, while the third column states most likely, but not assuredly, the number of polygon vertices.

Xml representation

Xml surpasses the object model by completely and relentlessly describing the information contained in the generalized matrix (1.3) as follows:

```
<polygon id="1" shape="triangle" vertices="3">
  <vertex1><x>0.0</x><y>0.0</y></vertex1>
  <vertex2><x>0.0</x><y>1.0</y></vertex2>
  <vertex3><x>1.0</x><y>0.0</y></vertex3>
</polygon>
```

```
<polygon id="2" shape="square" vertices="4">
  <vertex1><x>0.0</x><y>0.0</y></vertex1>
  <vertex2><x>1.0</x><y>0.0</y></vertex2>
  <vertex3><x>1.0</x><y>1.0</y></vertex3>
  <vertex4><x>0.0</x><y>1.0</y></vertex4>
</polygon>
```

It is fair to admit that the high density of this *xml* document is overwhelming. The generalized matrix (1.3) can be constructed from the *xml* data manually or automatically, but not *vice versa*.

DOM and SAX

Conversely, data encapsulated in an *xml* file can be imported and transformed into objects that can be manipulated or queried using an application written in a scientific language of our choice, such as *fortran* or C++. This methodology is the cornerstone of the document object model (DOM) where an *xml* file is mapped directly into a generalized object. The methods used to access an object are contained in the application programming interface (API).

The document object model (DOM) is useful for data sets with small or moderate size, but inappropriate for data contained in a large database. Direct query of the database using, for example, the simple API for *xml* (SAX) incorporating event handling callbacks is preferred in the second case.

Heterogeneous arrays and objects

If the data contained in a conceptual object are homogeneous (of the same data type), the object can be accommodated in a matrix (array) in *fortran* or C, subject to agreed conventions regarding the meaning of the columns or rows. Heterogeneous arrays are available in *perl* and other system programming languages. Arrays containing mixed data types can be stored as objects in object-oriented languages, such as C++.

1.6.3 Data points on a graph as xml elements or C++ objects

To illustrate the concept of objects, we consider data points in the xy plane representing a function to be plotted with colored symbols in a graph. Each point is defined by its x and y coordinates, color, and symbol type, such as a circle, square, or asterisk.

Xml elements

The data points can be conveniently recorded in the following complete *xml* file, including the *xml* declaration in the first line:

```
<?xml version="1.0"?>
<melomakarono>
```

```
  <datapoint>
    <x>0.0</x>
    <y>0.0</y>
    <color>black</color>
    <symbol>circle</symbol>
  </datapoint>

  <datapoint>
    <x>0.1</x>
    <y>0.2</y>
    <color>red</color>
    <symbol>asterisk</symbol>
  </datapoint>

</melomakarono>
```

The name of the root element is `melomakarono`. Two data points are defined in this file. Additional data points can be added following the chosen *xml* data tree.

C++ objects

The data points are now regarded as objects (members) of a class named *datapoint.* The following self-contained C++ code residing in a file named *datapoint.cc* defines the class of data points:

```
#include <iostream>
using namespace std;

/* ------ datapoint class definition ------ */

class datapoint
{
public:
  datapoint(); // default constructor of an object
  datapoint(float, float, string, string); // parametered constructor
  void print() const;
private:
  float x;
  float y;
  string color;
  string symbol;
};

/* ------ datapoint class implementation -------*/

datapoint::datapoint()
{
```

```
x = 0.0;
y = 0.0;
color = "black";
symbol = "circle";
}

datapoint::datapoint(float px, float py,
  string pcolor, string psymbol)
{
x = px;
y = py;
color = pcolor;
symbol = psymbol;
}

void datapoint::print() const
{
cout << x << " " << y << " " << color << " " << symbol << endl;
}

/* ------ main program -------*/

int main()
{
  datapoint A = datapoint();
  A.print();
  datapoint B = datapoint(0.1, 0.2, "red", "asterisk");
  B.print();
  return 0;
}
```

Readers who are not familiar with the C++ programming language can refer to Table 1.1 for miscellaneous explanations.

The code initially defines the class *datapoint* and declares three public interface member functions: a default constructor, a parametered constructor, and a print function of a member's attributes. Four private member attributes undisclosed to the main program are then declared. The *datapoint* class implementation follows the class definition.

The last part of the C++ code consists of the main program. For illustration, the main program defines datapoint A using the default constructor and datapoint B using the parametered constructor. The attributes of the first point are printed by the statement:

```
A.print()
```

The dot operation is commonplace in object-oriented programming.

#include <iostream>	Instructs the C++ preprocessor to attach a header file containing the definition, but not the implementation, of functions in the input/output stream library.
using namespace std;	The names of the functions defined in the standard `std` system library are adopted in the code.
public: private:	Member attributes are declared as *public* if available to the main program and functions of a different class, and *private* otherwise. Similarly, interface functions are declared as *public* if they can be called by the main program and functions of a different class, and *private* otherwise.
cout:	Internal library function for printing.

TABLE 1.1 Explanation of various lines in the C++ program *datapoint.cc* listed in the text containing information on data points.

Generating an executable

C++ compilers are included in standard *unix* distributions and are freely available on Windows.* To compile the *datapoints.cc* code and create an executable binary file named *datapoints*, we open a terminal (command-line window) and issue the statement:

```
c++ -o datapoints datapoints.cc
```

To run the executable, we type its name and press the ENTER key,

```
./datapoints
```

To ensure that the path of executables includes the current directory, we have inserted the dot-slash pair (./) in front of the name of the executable. The dot represents the current directory (folder) and the slash is a delimiter of the directory path. Running the executable prints on the screen:

```
0 0 black circle
0.1 0.2 red asterisk
```

It is clear that a C++ code is able to hold information on objects described as heterogeneous arrays. Descendant objects can be constructed as offsprings of parental objects using the concept of inheritance in object-oriented programming. These impressible features explain the phenomenal success of C++ and other object-oriented languages in applications programming.

*http://sourceforge.net/projects/mingw

Two practical questions naturally arise: how can we get a C++ code to print *xml* output in a way that both presents and describes the data? how can we import *xml* data into an C++ code? The answer to the first question is relatively straightforward. The answer to the second question is less straightforward, as discussed in Chapter 5.

1.6.4 Perl associative arrays

Perl is a powerful interpreted system programming language. The qualifier *system* emphasizes that the language is used mostly for retrieving and manipulating existing information, and to a lesser extent for generating new information. An outline of the basic language features is given in Appendix B. It is not necessary to compile a *perl* program, typically called a script, into an executable. The instructions contained in the script are executed as they are parsed by the *perl* interpreter.

Scalars and arrays

Perl allows us to use scalar variables, homogeneous arrays with uniform data types, and heterogeneous arrays with different data types, including integers, real number, and character strings. In this light, a *perl* array appears as an object described by numerical and narrative attributes. The value of a *perl* scalar variable is defined or extracted by prepending the dollar sign (`$`) to the variable name. The contents of a *perl* array are defined or extracted by prepending the *at* symbol (`@`) to the array name.

Hashes

A *perl* hash is a *perl* array endowed with references linking variable names (keys) to values that can be numbers or character strings. Thanks to the keys, a *perl* hash defines and describes in simple terms the data it encapsulates. A *perl* hash can be regarded as a map reminiscent of a dictionary. Accordingly, a *perl* hash is also called an associative array. To define or extract the contents of a hash, we prepend the percent symbol (`%`) to the hash name.

Data points

Each of the two data points defined in Section 1.6.3 can be accommodated into a *perl* hash, as shown in the following self-contained *perl* script residing in a file entitled *datapoints.pl*:

```
#!/usr/bin/perl

%datapoint1 = ( x => 0.0,
                y => 0.0,
                color => "black",
                symbol => "circle"
                );
```

```
%datapoint2 = ( x => 0.1,
                y => 0.2,
                color => "red",
                symbol => "asterisk"
                );

print "$datapoint1{color} \n";
print "$datapoint2{symbol} \n";
```

The first line identifies the directory where the *perl* interpreter resides in our *unix* system. One named *perl* hash is defined and evaluated for each data point. Note that each *perl* statement terminates with a semi-colon (`;`). The *perl* hashes defined in this script contain human-readable information for each data point.

The color of the first point is extracted as a scalar value (`$`) in the penultimate line of the script, and the symbol of the first point is extracted as another scalar value (`$`) in the last line of the script. A hash index analogous to a vector subscript is implemented by a pair of curly brackets (`{}`). The extracted variables are printed by two print statements in the last two lines. The character referenced by the `\n` pair forces a new line in the output at the end of each `print` statement.

Interpretation

Perl interpreters are included in standard *unix* distributions and can be obtained freely in other operating systems.* To execute a *perl* script, we open a terminal (command-line window) and type the name of the script followed by the ENTER keystroke:

```
./datapoints.pl
```

To ensure that the path of executables includes the current directory, we have inserted the dot-slash pair (`./`) in front of *perl* file name. Running the script produces the display:

```
black
asterisk
```

We see that a *perl* hash provides us with an attractive method of describing and defining simple objects.

Data points as a named array

The two data points under discussion, or any number of data points, can be ar-

*http://www.perl.org/get.html

ranged in a named array of anonymous *perl* hashes, called `datapoints`, defined as:

```
@datapoints = (
                {
                    x => "0.0",
                    y => "0.0",
                    color => "black",
                    symbol => "circle"
                },
                {
                    x => "0.0",
                    y => "0.1",
                    color => "red",
                    symbol => "asterisk"
                }
                );
```

Note that the array symbol (`@`) has been prepended to the array name to indicate array evaluation. Each component of this array is an anonymous hash enclosed by pairs of curly brackets (`{}`), accessible by an array index.

Conceptually, the data contained in this array can be accommodated in the rows of a generalized matrix,

$$\text{datapoints} = \begin{bmatrix} 0.0 & 0.0 & \text{black} & \text{circle} \\ 0.0 & 0.1 & \text{red} & \text{asterisk} \end{bmatrix},$$

where the first row receives the index 0 and the second row receives the index 1. In *perl*, as in C++, index counting begins at 0 so that all available bits of the integer counter are exploited, including a string of binary zeros.

To extract and print the properties of the first datapoint indexed 0, we use the lines:

```
print $datapoints[0]{x};
print $datapoints[0]{y};
print $datapoints[0]{color};
print $datapoints[0]{symbol};
```

Recall that the dollar sign (`$`) indicates a scalar. The screen display is:

```
0.0 0.0 black circle
```

To access and print the coordinates and properties of the second data point indexed 1, we replace `[0]` by `[1]` in the print statements. The complete code resides in the file *datapoints.pl* accompanying this book.

To illustrate the flexibility of *perl*, now we arrange the data into an anonymous array of anonymous hashes:

```
$tirith = [
          {
              x => "0.0",
              y => "0.2",
              color => "black",
                  symbol => "circle"
          },
          {
              x => "0.3",
              y => "0.1",
              color => "maroon",
              symbol => "diamond"
          }
        ];
```

The variable **tirith** is a scalar reference to an anonymous array enclosed by the square brackets ([]). The contents of the anonymous array are the same as those of the named array discussed previously.

To extract and print the properties of the first datapoint indexed 0, we use the lines:

```
print $tirith->[0]{x};
print $tirith->[0]{y};
print $tirith->[0]{color};
print $tirith->[0]{symbol};
```

The screen display is:

```
0.0 0.2 black circle
```

Note that the reference **tirith** is dereferenced by the *ascii* arrow consisting of two characters (->), as discussed in Appendix B.

A graph

Information on a complete graph of data points can be accommodated into an anonymous hash represented by a reference containing data points and other relevant information, defined as:

```
$graph = {
            datapoint => [
                          {
                              x => "1.0"
                             ,y => "3.0"
```

```
                     ,color => "black"
                     ,symbol => "circle"
                 },
                 {
                     x => "3.0"
                     ,y => "3.1"
                     ,color => "red"
                     ,symbol => "asterisk"
                 },
                     ]
        ,xlabel => "distance"
        ,ylabel => "temperature"
        ,title => "temperature distribution"
    };
```

The scalar variable **graph** is a reference to the outermost anonymous hash enclosed by the outermost curly brackets ({}). The scalar key **datapoint** inside the outer hash represents an anonymous array, indicated by the square brackets ([]), containing as elements anonymous hashes enclosed by the inner curly brackets ({}). The outermost anonymous hash contains three more keys defining the axes labels (`xlabel` and `ylabel`) and the graph title.

Appending to this script the lines:

```
print $graph->{datapoint}[1]{x};
print $graph->{datapoint}[1]{y};
print $graph->{datapoint}[1]{color};
print $graph->{datapoint}[1]{symbol};
print $graph->{xlabel};
print $graph->{ylabel};
print $graph->{title};
```

produces the screen display:

```
3.0 3.1 red asterisk distance temperature temperature distribution
```

The *ascii* arrow consisting of two characters (->) leads us from the reference to the content of the outermost anonymous hash.

Why not perl?

We have seen that a *perl* array of hashes can be used to *store and describe* data with inhomogeneous content, with the added advantage that the data can be manipulated using *perl* language instructions. It is clear that *perl*, or any other comparable language, such as *python*, is a viable alternative to *xml*.

Three main concerns in using *perl* and other similar system programming languages for data representation are: (*a*) the principle of code from data separation is likely to be violated, (*b*) difficulties in accommodating data with advanced structure may be encountered, and (*c*) computer programming experience is necessary. With regard to the third concern, we emphasize that an *xml* document can be written and edited by a person who is unfamiliar with any computer language. In practice, *perl* and similar high-level languages are used for *xml* data manipulation, as discussed in Chapter 5.

1.6.5 Computing environments

We have seen that the information encapsulated in an *xml* file can be arranged into homogeneous or inhomogeneous data structures of advanced programming languages, such as *perl*. Proprietary computing environments, such as *Matlab* and *Mathematica*, have made pertinent accommodations. For example, a *Matlab* structure can be defined using the statements:

```
student.name = 'Kathryne Marple';
student.gpa = '4.0';
```

An array of structures can be built into arrays and accessed by indices, as discussed previously in this section for *perl*. The *Mathematica* environment makes analogous accommodations.

1.6.6 Summary

Three main features of the *xml* framework are: (*a*) separation of data from code, (*b*) ability to collect, record, and retrieve data with a generic application in mind, and (*c*) lack of the requirement for computer programming skills. Specific data contained in an *xml* database can be extracted, mined or retrieved, imported, and manipulated by a person or program (application) with a particular goal in mind.

The two salient questions posed earlier in this chapter must be addressed: how can we get a computer code to generate and record *xml* output in a file? how can *xml* data be read efficiently from a code? An overview of available options will be given in this book.

Exercises

1.6.1 *Size of symbols*

Endow the data points defined in the C++ code discussed in the text with one additional attribute concerning the symbol size.

1.6.2 *Perl*

Write a *perl* script that describes two objects of your choice. Each object should be defined by a few alphanumerical properties (attributes).

Xml essential grammar 2

In Chapter 1, we explained the motivation behind *xml* data and statement formatting, outlined the basic structure of an *xml* document, and discussed possible applications. In this chapter, we summarize the basic *xml* grammar and illustrate the implementation of a document type definition (*dtd*) or *xml* schema definition (*xsd*) for the purpose of document validation. We recall that validation is necessary only when an *xml* document structure is required to conform with agreed conventions.

We have already emphasized that computer programming experience is not necessary for generating and editing an *xml* document containing data. Consequently, this chapter can be read and understood in its entirety by a person who has never written a computer code.

2.1 Xml tags

Xml grammar and syntax are sensible and intuitive. *Xml* tags are enclosed by pointy brackets, also called angle brackets,

```
< ··· >
```

where the three dots represent an uninterrupted string of appropriate characters defining the tag name, followed by optional attributes, as discussed in Section 2.2.1. For a document to be well-formed, an opening tag, such as

```
<dianxin>
```

must be succeeded by a corresponding closing tag indicated by a slash (/)

```
</dianxin>
```

where *dianxin* is an arbitrary tag name. Words, sentences, numbers, solitary tags, or nested tags can be enclosed between an opening tag and its closure. In fact, a whole book can be written inside a single tag and its closure.

Tag names

Xml tag notation is lower and upper-case sensitive. This means that a tag named *skordalia* is not the same as the tag *Skordalia*. Tag names must be uninterrupted, that is, they may not contain empty space. When an empty space is desirable, the underscore (_) should be employed as a compromise. Tag names may not begin with any of the following strings:

```
xml  Xml xML xmL XML XmL xML XML
```

In addition, tag names may not begin with numbers or contain any of the following characters:

```
; @ # $ % ^ ( ) % + ? =
```

Thus, the tag `<$bisque>` is not acceptable. The colon (`:`) is a special symbol reserved to indicating a namespace. The dash (`-`) and dot (`.`) characters should be avoided.

Self-closing tags

Self-closing tags can be employed. An example is the empty tag:

```
<nothing_to_see_here/>
```

This compact structure is equivalent to the verbose structure:

```
<nothing_to_see_here> </nothing_to_see_here>
```

In the framework of the *xsl* programming language discussed in Chapter 3, a self-closing tag may serve to launch an application. In typesetting a document, a self-closing tag may force a line break or start a new chapter.

Self-closing tags are not necessarily devoid of information. For example, a self-closing tag may introduce an element described by attributes residing next to the tag name, as discussed in Section 2.2.1. An example is the tag:

```
<change_color new_color="moccasin"/>
```

where `new_color` is an attribute evaluated as *moccasin*.

Textual content

Text consisting of individual characters, words, and numbers can be inserted between an *xml* tag and its closure. Words and numbers broken into pieces by empty spaces lose their wholesome meaning and are treated as separate entities. For example, it is not appropriate to write:

```
<pi>3.14159 265358</pi>
```

The textual content of this element will be interpreted as a character string involving a blank space, not as a number. However, statements are allowed to extend over an arbitrary number of lines and the invisible character forcing a line break is inconsequential. A continuation mark at the end of a line is not required. For example, we may write:

```
<mytoolbox> currently, the toolbox is empty;
           please send in donations (especially hammers).</mytoolbox>
```

Exercise

2.1.1 *Self-closing tag*

Provide a sensible example of a self-closing tag involving an attribute.

2.2 Xml elements

Anything enclosed between a tag and its closure is an *xml* element, also called an element node. An element can be a physical object, an abstract object, a property of a parental object, or an instruction of a computer language that conforms with *xml* syntax and grammar. A self-closing tag is a vacant element.

2.2.1 Element attributes

An *xml* element may have attributes with arbitrary names conveying properties or descriptions, as illustrated in the following example:

```
<car color="red with black seats">
  ...
</car>
```

In this case, `color` is an attribute of a `car` evaluated by the character string `red with black seats`. The three dots denote additional content describing further element properties.

When employed, an *xml* attribute must be evaluated. For example, it is *not acceptable* to state:

```
✗ THIS IS WRONG:

<car lemon>
  ...
</car>
```

The value of an attribute, whether a number or character string, must be enclosed by single or double quotes. For example, we may write:

```
<broom id="1" type= 'straw' color= "red"/>
  ...
</broom>
```

Element attributes can be used in self-closing tags, as discussed in Section 2.1. For example, we may write:

```
<force type="gravitational"/>
<force type="electromagnetic"/>
<force type="Coriolis"/>
<force type="centrifugal"/>
```

It is a good practice to avoid using attributes as much as possible and employ element nesting to describe object properties instead, as discussed in Section 2.2.2. One reason is that extracting attribute values requires more elaborate code. Another reason is that attributes cannot grow into data trees, and this may necessitate the restructuring of an *xml* document when additional information is supplied. This observation underlines the importance of proactive *xml* document design.

2.2.2 Property listing and nesting

Pairs of tags representing elements may be listed sequentially or otherwise nested multiple times to define a hierarchy of substructures. For example, we may write:

```
<car color="red with black seats">
  <make>Wartburg</make>
  <year>2007</year>
</car>
```

The word *Wartburg* should be regarded as the textual content, not the value, of the `make` element inside the `car` element. Similarly, we may write:

```
<car>
  <new>
    <make>Wartburg</make>
    <year>2007</year>
  </new>
</car>
```

The number *2007* should be regarded as the textual content, not the value, of the `year` element inside the `new` element.

Pairs of tags may not cross-over or overlap. Thus, the following structure is *not* acceptable:

```
✗ THIS IS WRONG:

<polynomial><orthogonal>Legendre</polynomial></orthogonal>
```

The correct structure is:

```
<polynomial><orthogonal>Legendre</orthogonal></polynomial>
```

Although *xml* tags can be arranged in a single line to save line breaks, this obscures the element structure.

The following *xml* element describes a triangle in terms of the coordinates of the three vertices in the xy plane specified as different *xml* children elements of the triangle:

```
<triangle>
  <x1>0.0</x1> <y1>0.0</y1>
  <x2>0.5</x2> <y2>0.3</y2>
  <x3>0.6</x2> <y3>-0.1</y3>
</triangle>
```

A person with elementary knowledge of geometry should be able to draw the triangle. If the order of the vertices is irrelevant and inconsequential, the triangle could be described as:

```
<triangle>
  <vertex> <x>0.0</x> <y>0.0</y> </vertex>
  <vertex> <x>0.5</x> <y>0.3</y> </vertex>
  <vertex> <x>0.6</x> <y>-0.1</y> </vertex>
</triangle>
```

This example illustrates that multiple vertex elements populating the same triangle element are allowed. To order the vertices, we may use an attribute:

```
<triangle>
  <vertex order="3'> <x>0.0</x> <y> 0.0</y> </vertex>
  <vertex order="1'> <x>0.5</x> <y> 0.3</y> </vertex>
  <vertex order="2'> <x>0.6</x> <y>-0.1</y> </vertex>
</triangle>
```

The `order` attribute allows us to assess whether the three vertices are arranged in the counterclockwise fashion in the xy plane by performing an appropriate geometrical test.

Talking out of turn

If we had misprinted the order of a vertex of a triangle so that two different vertices have the same order, the *xml* document would still be well-formed. This observation indicates that conforming with *xml* grammar does not guarantee contextual or mathematical sense.

Mixed content

Consider the following *xml* element:

```
<polynomial>
  Legendre
    <degree>23</degree>
</polynomial>
```

This element has mixed content consisting of (a) character data spelling the name *Legendre* and (b) one nested child element specifying the degree of the polynomial.

It is a good practice to avoid using mixed content as much as possible. When *xml* data are processed by an application written in the *xsl* language, mixed data are typically handled by templates playing the role of functions or subroutines, as discussed in Chapter 3.

2.2.3 Property and element tag names

Two different property tags may have the same name, provided that the corresponding elements are uniquely identified. For example, the following name scheme can be employed:

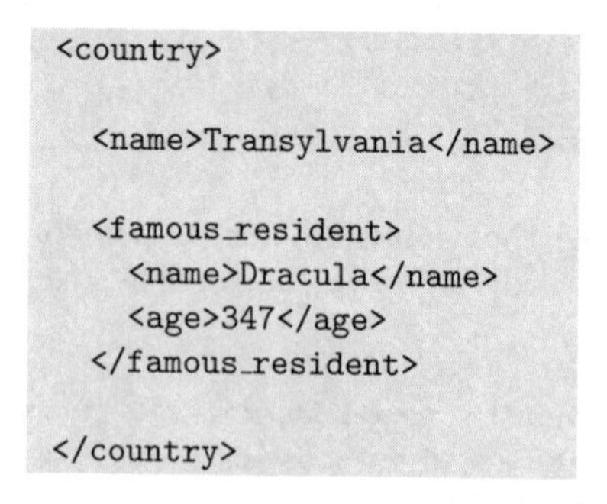

```
<country>

  <name>Transylvania</name>

  <famous_resident>
    <name>Dracula</name>
    <age>347</age>
  </famous_resident>

</country>
```

Note that the `name` tag appears twice with different meanings in this data structure. This is perfectly acceptable, for the context in which each name appears is clear.

White space

White space is generated by pressing the space bar, the TAB key, the ENTER or RETURN key. *Xml* parsers are trained to retain white space inside an element, ignore white space between elements, and normalize white space by condensing it into a single space in attribute evaluations.

Exercises

2.2.1 *Record a truck and then a binomial*

(*a*) Record a truck as an *xml* element described by its make, year, color, and number of doors. (*b*) Record a binomial, $ax^2 + bx + c$, described by three possibly complex coefficients, a, b, and c.

2.2.2 *Mixed content*

Discuss a case where an *xml* element can be sensibly endowed with mixed content.

2.3 Comments

Comments are extremely helpful for providing explanations, documentation, and ancillary information in a data file or computer code. Comments can be inserted anywhere in an *xml* file according to the following format:

```
<!-- This was written on February 29, 2012 -->
```

or

```
<!-- The test of a first-rate intelligence is the ability
to hold two opposed ideas in the mind at the same time
and still retain the ability to function.

                F. Scott Fitzgerald -->
```

Xml parsers are trained to ignore material between the comment delimiters `<!--` and `-->`. The *xml* comment convention is the same as that used in *html*. *Latex* accepts comments indicated by the percent mark (%) at the beginning of each commenting line or at any place in partially commenting line.

Two consecutive dashes (--) may not appear inside an *xml* comment.* For example, the following comment is not permissible:

```
<!-- use the Crank--Nicolson method -->
```

Although a comment may contain *xml* elements, an *xml* tag may not contain comments. Comments can be inserted inside the document type declaration DOCTYPE block defining a document type definition (*dtd*), as discussed in Section 2.10.

*A double dash encodes an *en dash* in *latex* typesetting, separating words that could each stand alone. In contrast, the hyphen separates words that convey meaning only as a pair. Thus, we must write: The Navier–Stokes equation is a second-order differential equation in space.

Commenting out blocks

The comment delimiters can be used to softly remove individual elements or groups of elements in an *xml* document. In the following example, delimiters are used to remove two zeros of a Bessel function:

```
<bessel_J0>
  <root>2.4048</root>

<!--
  <root>5.5201</root>   <root>8.6537</root>
-->

</bessel_J0>
```

Why comment out instead of remove? The discarded material may need to be temporarily disabled for a variety of reasons. Calmly think of a telemarketer removing a telephone number after receiving a complaint, only to reinstate it at a later time.

In scientific programming, commenting out lines is an invaluable method of debugging code. To comment out a line in *fortran*, we insert the `c` character at the beginning of the line. To comment out the whole or the tail end of a line in *fortran*, we put an exclamation mark (`!`) anywhere in the line. Text enclosed by the begin doublet `/*` and the end doublet `*/` is ignored in C or C++ code. To comment out the whole or tail end of a line in *Matlab*, we put a percent sign (`%`) anywhere in the line.

To comment out a block of text in a *latex* document, we use the *verbatim* package and wrap the disabled text inside the *comment* tag,

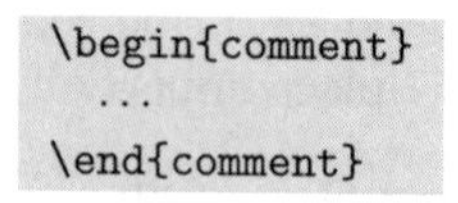

```
\begin{comment}
  ...
\end{comment}
```

where the three dots represent deactivated text.

Exercise

2.3.1 *Triple dash*

Can we put a triple dash (`---`) inside an *xml* comment?

2.4 Xml document declaration

The first line in an *xml* file declares that the file contains an *xml* document consistent with a specified *xml* version and possibly with a chosen character set. The minimal declaration stating consistency with the *xml* version 1.0 specification is:

```
<?xml version="1.0"?>
```

Version 1.1 extends the range of characters that can be employed.

Character sets

The *unicode* is a protocol mapping over one million characters, including letters, numbers, and other symbols, to a set of integers ranging from 0 to 1, 114, 112. Two mapping methods are available: the unicode transformation format (UTF) and the universal character set (UCS).

The UTF-8 and UTF-16 encodings are commonly employed.* UTF-8 maps characters to integers represented by 8 bits (one byte), whereas UTF-16 maps characters to integers represented by 16 bits (two bytes). The UTF-8 set is most compatible with the legacy American Standard Code for Information Interchange (*ascii*) mapping listed in Appendix A.

If western European languages are only used, the single byte ISO-8859-XX character set can be adopted, where `xx` is the version number. The ISO-8859-1 set is used by default in documents whose media type is *text*, such as those handled by a *web* server.

Character encoding and standalone specification

A typical *xml* declaration stating consistency with *xml* version 1.0 and the use of the ISO-8859-1 character set is:

```
<?xml version="1.0" encoding="ISO-8859-1"?>
```

The *xml* data and possible processing instructions follow this declaration, as discussed later in this section. The most general *xml* document declaration has the typical form:

```
<?xml version="1.0" encoding="ISO-8859-1" standalone="yo"?>
```

where `yo` can be `yes` or `no`. The optional `standalone` attribute is used when an internal document type definition (*dtd*) is employed, as discussed in Section 2.10.

It is worth emphasizing that the *xml* declaration becomes relevant only when an *xml* document is supplied for processing to an application. Particular applications may demand specific character sets.

---

*http://www.iana.org/assignments/charset-reg

Exercise

2.4.1 *ISO-8859-1*

List the first sixteen characters implemented in the ISO-8859-1 character set.

2.5 Character reference

Instead of typing a character, such as the pound sign (`#`), we may reference its *unicode*. For example, if the numerical code of a character in the decimal system is `78`, we may reference the character by entering:

```
&#78;
```

Notice the mandatory semi-colon (`;`) at the end. If the numerical code of a character in the hexadecimal system is `B6`, we may reference the character by entering:

```
&#xB6;
```

The hexadecimal encoding is indicated by the character `x`.

Predefined entities

Several predefined entities are available in *xml*. The *less than* (`<`) and *greater than* (`>`) signs, recognized as pointy or angle brackets and used as *xml* tag containers, can be referenced as:

```
&lt;     &gt;
```

The ampersand (`&`) character can be referenced as:

```
&amp;
```

For example, we may state:

```
<spice>
  salt &amp; pepper
</spice>
```

After the text enclosed by the spice tags has been processed *xml* processor, the following text will appear in the output: *salt & pepper*. Other predefined entities include the double quotation mark (") referenced as

```
&quot;
```

and the apostrophe or single quotation mark (') referenced as

```
&apos;
```

Exercise

2.5.1 *A word by character reference*

(*a*) Record the word *sanctimonious* by character reference to the Unicode. (*b*) Repeat for the word *promulgate*.

2.6 Language processing instructions

An instruction implemented in an appropriate language, such as *xsl*, *perl*, *python*, *java*, and others, or application, is called a processing instruction (PI). To include a processing instruction in an *xml* document, we place it in a container opening with the pair `<?` and closing with the mirroring pair `?>`. A processing instruction becomes relevant only at the stage where an *xml* document is supplied to a processor for manipulation, bearing no relevance to the structure of the *xml* document.

Extensible stylesheet (xsl)

For example, an *xml* document may contain the following processing instruction whose meaning will be discussed in Chapter 3 in the context of the *xsl* processor:

```
<? xml-stylesheet type="text/xsl" href="bilo.xsl" ?>
```

Briefly, this line invokes an extensible stylesheet (*xsl*) residing in a file whose name (*bilo.xsl*) is provided as a hypertext reference (*href*) attribute.

Cascading stylesheet (css)

Processing instructions are used routinely to link an *xml* or *html* document to a cascading stylesheet (*css*) by the typical statement:

```
<? xml-stylesheet href="mystyle.css" type="text/css" ?>
```

A cascading stylesheet defines the global formatting of typesetting elements in an *xml* or *html* file. For example, a *css* may define the default display font, as discussed in Section 3.13.

Proprietary and other applications

Processing instructions can be used to convey information to a proprietary application, such as a word processor or a spreadsheet. A typical usage is:

```
<? mso-application progid="Excel.Sheet" ?>
```

The name of the application (*Excel.Sheet*) is provided as a program id (*progid*) attribute.

A processing instruction can be used to insert comments in an *xml* document. However, better methods of inserting comments are available.

Processing instructions cannot be placed inside processing instructions, that is, they cannot be nested.

Xml declaration

In spite of its deceiving appearance, the *xml* declaration at the beginning of an *xml* document, such as

```
<?xml version="1.0" encoding="ISO-8859-1"?>
```

is *not* a processing instruction. The reason is that this declaration is completely understood by an *xml* parser. In contrast, a PI can be understood only by an external application parsing the *xml* file.

Exercise

2.6.1 *Use of PIs*

Investigate the use of a PI in an application of your choice.

2.7 Character data (CDATA)

Suppose that an *xml* document contains text that includes the following line:

```
<eigenvalue>0.302</eigenvalue>
```

which is to interpreted as verbatim text, as opposed to an *xml* element. Unfortunately, the string will be misconstrued as an *xml* element by the *xml* parser.

To prevent this misinterpretation, we may recast the line in terms of its character components as:

```
&lt; eigen &gt; 0.302 &lt; /eigenvalue&gt;
```

which eliminates the explicit presence of the troublesome pointy brackets (<>).

A more elegant and less confusing method involves putting the verbatim text inside the structure:

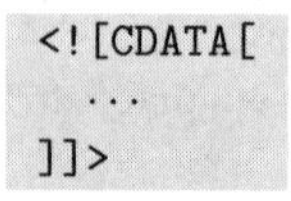

```
<![CDATA[
  ...
]]>
```

where the three dots indicate arbitrary text and CDATA stands for *character data* to be ignored by the *xml* parser. Note that the keyword CDATA must be capitalized. Also note the presence of two nested square brackets. In our example, we write:

```
<![CDATA[ <eigenvalue>0.302</eigenvalue> ]]>
```

In *web* programming applications, the character data structure is often used to enclose code.

It is understandable why unparsed character data (CDATA) may not contain the sequence:

```
]]>
```

The reason is that this pair will be falsely interpreted as the closing CDATA delimiter.

Deprecated tags allow us to insert verbatim text in an *html* document. Verbatim text in *latex*, such as #$%(*&&, can be placed inside the verbatim environment:

```
\begin{verbatim}
  #$%(*&&
\end{verbatim}
```

Verbatim text with small length can be placed inline using the typical *latex* structure:

```
\verb:this text appears verbatim:
```

where the semicolon (:) can be replaced by another character.

CDATA and PCDATA

To be precise, character data (CDATA) should be called unparsed character data, meaning that they are not parsed by an *xml* processor for the purpose of identifying elements and other structural information. In contrast, parsed character data are denoted as PCDATA.

Exercise

2.7.1 *Text explaining CDATA*

Is it possible to write a sentence discussing the CDATA statement in an *xml* document?

2.8 Xml root element

An *xml* file *must* contain a root element that encloses data and statements to be read by a person or processed by an application. Sometimes the root element is called the *document.*

The first tag in an *xml* file following the *xml* declaration and possible processing instructions defines the root element, and the last tag defines the closure of the root element. The typical structure of an *xml* document is:

```
<?xml version="1.0" encoding="ISO-8859-1"?>
<root_element_name>
  ...
</root_element_name>
```

where `root_element_name` can be any suitable name, and the three dots indicate additional data. The root element can be endowed with attributes. The root element of an *xhtml* document is `<html>`, and its closure is `</html>`. All elements inside the root element are children or descendants of the root element.

Self-closing root element

Strange though it may appear, a self-closing root element could be meaningful. For example, the following *xml* file may serve a purpose:

```
<?xml version="1.0" encoding="ISO-8859-1"?>

<shutdown what="system" when="now"/>
```

The name of the root element defined in the second line is `shutdown`. The mere presence of the root element is capable of triggering a specific type of action when the *xml* document is parsed to be processed by an application.

DOCTYPE declaration

The name of the root element in an *xml* document can be stated explicitly in the preamble by the statement:

```
<!DOCTYPE root_element_name>
```

More generally, the DOCTYPE declaration defines a data type definition (*dtd*), as discussed in Section 2.10.

Only one root element may be present in an xml file

Under no circumstances an *xml* file may have two root elements. Thus, the following structure is *not acceptable:*

```
✗ THIS IS WRONG:

<?xml version="1.0" encoding="ISO-8859-1"?>
<canoli>
  ...
</canoli>

<baklavas>
  ...
</baklavas>
```

One must choose either `canoli` or `baklavas`; it is wrong to indulge in both.

The root element of an xml data file is not a main program

Scientific computer programmers may be tempted to make a correspondence between the root element of the *xml* document containing data, the main program of a *fortran* code, or the main function of a C or C++ code. However, this correspondence is false. The sole similarity is that an *xml* data file may have one root element, and a C or C++ code may have only one main function.

In the *xml/xsl* framework discussed in Chapters 3 and 4, the root element of an *xml* document triggers the execution of the *xsl* code. The notion of data driving the execution is foreign to scientific programmers who are used to regarding data as optional companions of a standalone code.

The root element of an xml program file is not a main program

An *xml* file may contain code implementing computer language instructions, as discussed in Section 2.12. Even in these cases, the root element is not a main program or function, but only serves to introduce and set up the language processor, as discussed in Chapters 3 and 4 in the *xml/xsl* framework.

Exercise

2.8.1 *Unix root directory*

Discuss the analogy, if any, between the root directory of a *unix* system and the root element of an *xml* document.

2.9 Xml trees and nodes

Element nesting in an *xml* document results in an *xml* tree originating from the highest branching point called the *root*. Consider the following *xml* document:

```
<?xml version="1.0" encoding="ISO-8859-1"?>
<equation>
```

```
<algebraic>
  <quadratic>
    <first_coefficient> 1.5 </first_coefficient>
    <second_coefficient> 3.0 </first_coefficient>
    <third_coefficient> -4.0 </third_coefficient>
  </quadratic>
  <cubic>
    ...
  </cubic>
</algebraic>

<differential>
  ...
</differential>

<integral>
  ...
</integral>

</equation>
```

where three dots indicate additional lines of data. The name of the root element, enclosing all other children elements, is `equation`. Different types of equations are recorded in this document according to their classification.

The *xml* element structure forms a tree of nodes originating from the root, as depicted in Figure 2.1. Each labelled entry in this tree is an element node. Sibling, ascendant, and descendant nodes can be identified in an *xml* tree. In our example, algebraic, differential, and integral equations are siblings. Each node has one only one parent node. A leaf is a childless node.

Node list

Different quadratic algebraic equations defined in the document under discussion constitute a *node list* populating the same element node. Each quadratic algebraic equation could be attached to the appropriate branch near the south-western portion of the tree depicted in Figure 2.1, labeled by a numerical index starting at 0. Any *xml* element node can be populated with an arbitrary number of children nodes forming a node list parametrized by a numerical index.

Navigation path

It is important to emphasize that, even though the same element name may appear twice or multiple times at different branches of an *xml* tree, the corresponding nodes are distinguished by separate navigation paths. In our example, two such paths leading to an element named `second_order` are:

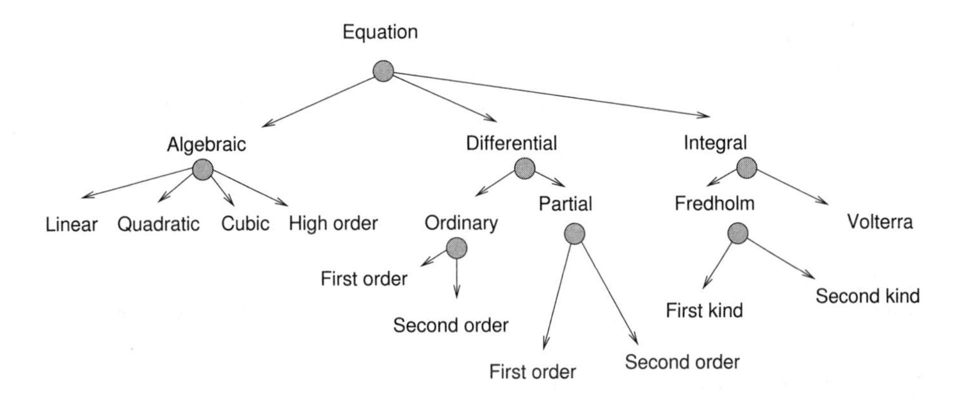

FIGURE 2.1 Tree structure of an *xml* document describing equations. The circles represent document nodes possibly serving multiple customers.

```
equation -> differential -> ordinary -> second_order
equation -> differential -> partial  -> second_order
```

These paths are analogous to library or executable paths of an operating system (Os). In an *xml* document, each path defines a unique *xml* node.

Relations and design

Relational context and possible future extensions must be understood for a successful data organization in an *xml* tree. Unless a large amount of disparate data are involved, common sense and a basic understanding of concepts and entities described in the *xml* document are the only prerequisites.

Basic rules

Two basic rules for an *xml* document to be well-formed may now be identified: (*a*) all tags must be properly nested and (*b*) only one root element playing the role of a main program may be present.

To illustrate further the meaning of proper nesting, we consider the following sentence: *I entered the barn, fed the donkey, exited the barn, and pet the donkey.* This sentence is not well-formed in the *xml* or any other rational framework. The proper structure is: *I entered the barn, fed the donkey, pet the donkey, and exited the barn.* In *fortran*, we can have an `If` loop inside a `Do` loop, but the `If` loop must close with an `End If` statement before the `Do` loop closes with an `End Do` statement.

Xml nodes

We have referred to the components of the tree shown in Figure 2.1 as element nodes. In fact, each identifiable component of an *xml* document is also an *xml* node. Examples include the root element (document), any other child or descendant element, an element attribute, a processing instruction (PI), or a document type definition.

Exercise

2.9.1 *Xml tree*

Draw an *xml* tree containing square matrices in some rational taxonomy.

2.10 Document type definition and schema

We have seen that an *xml* document can be written using tags of our choice for clarity and easy reference. Eventually, the data will be read by a person or processed by a machine running an application. To prevent misinterpretation and ensure that the data are complete, an agreement on the tagging system describing element properties and attributes must be reached.

2.10.1 Internal document type definition (dtd)

In the simplest method, the agreement is implemented in a document type definition (*dtd*) that can be part of an *xml* document or accompany an *xml* document in an external file.

Consider the following well-formed *xml* document contained in the file *vehicles.xsl* of interest to a used-car dealer:

```
<?xml version="1.0" encoding="ISO-8859-1"?>

<!DOCTYPE inventory [
  <!ELEMENT inventory (car|truck)* >
  <!ELEMENT car (make,year) >
  <!ELEMENT truck (factory,built) >
  <!ELEMENT make (#PCDATA) >
  <!ELEMENT year (#PCDATA) >
  <!ELEMENT factory (#PCDATA) >
  <!ELEMENT built (#PCDATA) >
  <!ATTLIST car color CDATA "black" >
  <!ATTLIST truck color CDATA "black" >
  <!ENTITY found "We found a " >
]>

<inventory>

<car color="red">
  <make>Wartburg</make>
  <year>1966</year>
```

Symbol	Meaning
|	The vertical bar stands for logical OR to allow for a choice.
+	An element becomes mandatory by appending a plus sign to its name in the *dtd*.
*	An element is rendered optional by appending an asterisk to its name in the *dtd*.
?	An element is rendered optional and multivalued by appending a question mark to its name in the *dtd*.
CDATA	CDATA stands for unparsed character data.
"	The quotes enclose default attributes.
#PCDATA	#PCDATA stands for parsed character data.

TABLE 2.1 Conventions employed in defining elements and attributes in a document type definition (*dtd*).

```
</car>

<car color="green">
  <make>Yugo</make>
  <year>1970</year>
</car>

<truck color="white">
  <factory>Mercedes</factory>
  <built>1944</built>
</truck>

</inventory>
```

The name of the root element is `inventory`. Each car or truck is described by one attribute and two properties implemented as nested *xml* nodes.

DOCTYPE declaration

An internal *dtd* referring to the root element of the *xml* file and its descendants is implemented inside the DOCTYPE declaration, following the *xml* declaration in the first line of the *xml* document. Note that the keyword DOCTYPE is printed in upper-case letters. The *dtd* is implemented before the root element of the *xml* file.

The first entry in the *dtd* defines the root element of the *xml* file. Subsequent entries define elements by the keyword !ELEMENT and element attributes by the keyword !ATTLIST, subject to the conventions shown in Table 2.1.

In our example, the internal *dtd* specifies that the make and then the year of each car or truck must be declared, consistent with the *xml* data following the *dtd*. The tags `make` and `year` define two children elements of cars, whereas the tags `factory` and `built` define two children elements of trucks. These doublets convey similar information on the builder and date of built of each vehicle.

Unfortunately, a *dtd* does not allow us to assign the same name to children elements of two different elements. An *xml* schema definition (*xsd*) must be used when assigning the same name is desirable or necessary, as discussed in Section 2.10.3.

Element declaration

The most general element definition in a *dtd* is:

```
<!ELEMENT element_name element_content>
```

where `element_name` is the given element name. The `element_content` block defines the children elements, as shown in the *vehicles.xml* file. Additional examples are shown in Table 2.2(*a*). Other choices for `element_content` include EMPTY, ANY, and a combination of parsed character data (#PCDATA) and children elements.

Element attribute declaration

The most general definition of an element attribute in a *dtd* is:

```
<!ATTLIST element_name attribute_name attribute_type attribute_value>
```

where `attribute_name` is the given attribute name describing the `element_name`.

The most common `attribute_type` is CDATA, as shown in the *vehicles.xml* file. Other choices, such as `ID`, are available. The `attribute_value` can be one of the following:

```
#REQUIRED      #IMPLIED      #FIXED "somevalue"      "somevalue"
```

The keyword IMPLIED is used for an optional attribute that does not have a default value. The keyword FIXED is used for a mandatory attribute whose value cannot be changed in the *xml* document. Examples are shown in Table 2.2(*b*).

Entities

An *entity* can be defined in a *dtd* and then referenced in the *xml* file. For example, an entity named `showme` can be defined as:

```
<!ENTITY showme " The temperature measured at this point is:  ">
```

(*a*)

<table>
<tr><td><!ELEMENT inventory (car|truck)></td><td>One car or one truck is expected.</td></tr>
<tr><td><!ELEMENT inventory (car|truck)*></td><td>Any number of cars and trucks can be interspersed.</td></tr>
<tr><td><!ELEMENT inventory (car*|truck*)></td><td>Any number of cars or trucks are allowed; cars and trucks are not allowed.</td></tr>
<tr><td><!ELEMENT body (car*, truck*)></td><td>Any number of cars may be followed by any number of trucks.</td></tr>
<tr><td><!ELEMENT body (car, truck)*></td><td>Any number of ordered pairs of cars and trucks are allowed.</td></tr>
<tr><td><!ELEMENT car (make, year?)></td><td>The make of a car must be stated before the year, but the year is optional.</td></tr>
</table>

(*b*)

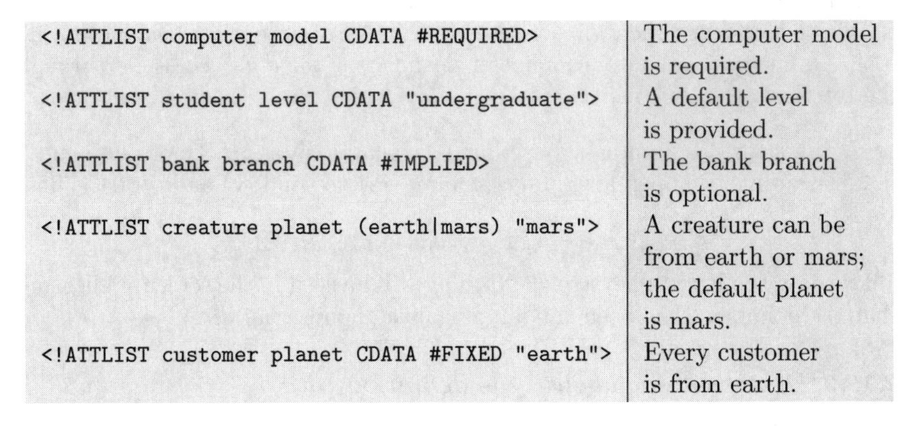

<table>
<tr><td><!ATTLIST computer model CDATA #REQUIRED></td><td>The computer model is required.</td></tr>
<tr><td><!ATTLIST student level CDATA "undergraduate"></td><td>A default level is provided.</td></tr>
<tr><td><!ATTLIST bank branch CDATA #IMPLIED></td><td>The bank branch is optional.</td></tr>
<tr><td><!ATTLIST creature planet (earth|mars) "mars"></td><td>A creature can be from earth or mars; the default planet is mars.</td></tr>
<tr><td><!ATTLIST customer planet CDATA #FIXED "earth"></td><td>Every customer is from earth.</td></tr>
</table>

TABLE 2.2 An assortment of (*a*) element definitions and (*b*) element attribute definitions.

In the *xml* file, this entity is referenced as

```
&showme;
```

For example, we may state:

```
<temperature> &showme; 100.0 </temperature>
```

Summary

An internal *dtd* is placed immediately after the *xml* document declaration. The general statement of an internal *dtd* is:

```
<!DOCTYPE name_of_the_root_element [
  ...
]>
```

where the three dots denote statements that define elements, element attributes, entities, notation, processing instructions, comments, and references.

Validation

The process of inspecting an *xml* document against a *dtd* is called validation. Validation can be performed by opening an *xml* file with a *web* browser that is able to perform validation, or else by using an appropriate *xml* authoring tool. Online *xml* validators for internal and external *dtd*s accessible through the Internet are available.

A validator program called *xmllint* is available on a variety of platforms.* In our example, we open a terminal (command-line window) and type the line:

```
xmllint --valid --noout vehicles.xml
```

followed by the ENTER keystroke. Nothing will be seen in the screen, indicating that the document has been validated against the internal *dtd*.

2.10.2 External document type definition (dtd)

A *dtd* can be placed in a separate file that accompanies an *xml* document. In that case, the internal *dtd* at the beginning of the *xml* file is replaced with the single line:

```
<!DOCTYPE inventory SYSTEM "DTD_file_name">
```

where `DTD_file_name` is the name of the file where the *dtd* is defined. The name of the *dtd* file is possibly preceded by a directory path, or else by a suitable

*http://xmlsoft.org

web address identified as a uniform resource locator (*url*). In the last case, the keyword SYSTEM in the *dtd* declaration can be replaced by the keyword PUBLIC. The external *dtd* file itself contains the text enclosed by the square brackets of an internal *dtd*.

If the PUBLIC keyword is used, a formal public identifier (*fpi*) must be used in the DOCTYPE declaration. The *fpi* consists of four fields separated by a double slash. For example, the DOCBOOK *dtd* discussed in Section 1.4.1 is invoked by the following statement where the *fpi* is printed in the second line:

```
<!DOCTYPE article PUBLIC
  "-//OASIS//DTD DocBook XML V4.1.2//EN"
    "http://www.oasis-open.org/docbook/xml/4.1.2/docbookx.dtd">
```

The dash in the first field of this *fpi* indicates that the *dtd* has not been approved by a recognized authority; the second field indicates that the organization OASIS is responsible for this *dtd*; the third field contains additional information; the fourth field is an English language specification (EN).

If an external *dtd* is used, the *xml* document declaration must specify that the *xml* document is not standalone. For example we may declare:

```
<?xml version="1.0" encoding="UTF-8" standalone="no"?>
```

If the `standalone` attribute is omitted, the default state is affirmative.

A graph

In mathematics, a graph is a set of nodes (vertices) connected by edges (links). Vertices and edges are assigned arbitrary and independent numerical labels. As an example, we consider the following file named *graph.xml* containing information on three vertices and two edges of a graph:

```
<?xml version="1.0" encoding="ISO-8859-1" standalone="no"?>

<!DOCTYPE network SYSTEM "graph.dtd">

<network>

<vertex id="1">
  <adjacent>2</adjacent>
  <adjacent>3</adjacent>
</vertex>

<vertex id="2">
  <adjacent>3</adjacent>
  <adjacent>1</adjacent>
</vertex>
```

```
<vertex id="3">
  <adjacent>1</adjacent>
  <adjacent>2</adjacent>
</vertex>

<edge id="1">
  <incidence1>1</incidence1>
  <incidence2>2</incidence2>
</edge>

<edge id="2">
  <incidence1>2</incidence1>
  <incidence2>3</incidence2>
</edge>

</network>
```

The name of the root element is `network`. The label of each vertex or edge is recorded as an attribute named `id`. The vertices adjacent to each vertex are recorded along with two incidence indices for each edge specifying the labels of the first and second end points of each edge. Cursory inspection reveals that this is a `V`-shaped graph consisting of three vertices connected by two edges.

The accompanying external *dtd* file named *graph.dtd* referenced in the second line reads:

```
<!ELEMENT network (vertex*,edge*) >
<!ELEMENT vertex (adjacent*) >
<!ELEMENT adjacent (#PCDATA) >
<!ELEMENT edge(incidence1, incidence2) >
<!ELEMENT incidence1 (#PCDATA) >
<!ELEMENT incidence2 (#PCDATA) >
<!ATTLIST vertex id CDATA #REQUIRED >
<!ATTLIST edge id CDATA #REQUIRED >
```

Assuming that the *graph.xml* and *graph.dtd* files reside in the same directory (folder), we may validate the *xml* data using the *xmllint* application by opening a terminal (command-line window) and issuing the command:

```
xmllint -noout graph.xml --dtdvalid graph.dtd
```

Nothing will appear on the screen, indicating that the *xml* file has been validated against the external *dtd*.

Combining an internal with an external dtd

An *xml* document can have an internal *dtd*, an external *dtd*, or both. In our example, we may use the declaration:

```
<!DOCTYPE network SYSTEM "graph.dtd" [
  <!ENTITY found "Number of nodes:  " >
]>
```

which adds an entity to the external *dtd*. The same element cannot be defined both in the internal and external *dtd*.

2.10.3 Xml schema definition (xsd)

Like a *dtd*, an *xml* schema definition (*xsd*) determines the required structure of an *xml* document.* The word schema (pl. schemata) should not be confused with the possibly pejorative scheme.

Unlike a *dtd*, an *xsd* may incorporate advanced features that allow us to define data types, such as integers, real numbers, and character strings, and also employ namespaces, as discussed in Section 2.11. The use of a *xsd* is recommended over a *dtd* in advanced and commercial applications.

An *xml* schema definition is contained in a file identified by the suffix `.xsd`. An interesting feature of an *xsd* is that its implementation follows *xml* grammar. As an example, we consider the data contained in the following *xml* file named *pets.xml*:

```
<?xml version="1.0" encoding="ISO-8859-1"?>
<pets xmlns:xsi="http://www.w3.org/2001/XMLSchema-instance"
    xsi:noNamespaceSchemaLocation="pets.xsd">

    <dog>Pluto</dog>
    <cat>Garfield</cat>

</pets>
```

The second line makes reference to a schema contained in the following file named *pets.xsd*:

```
<?xml version="1.0" encoding="UTF-8"?>
<xs:schema xmlns:xs="http://www.w3.org/2001/XMLSchema">

  <xs:element name="pets">
    <xs:complexType>
      <xs:sequence>
        <xs:element name="dog" type="xs:string" />
        <xs:element name="cat" type="xs:string" />
      </xs:sequence>
```

*$\Sigma\chi\eta\mu\alpha$ is a Greek word meaning shape, form, layout, or framework.

```
    </xs:complexType>
  </xs:element>

</xs:schema>
```

Assuming that the files *graph.xml* and *graph.xsd* reside in the same directory (folder), we may validate the *xml* data using the *xmllint* application by opening a terminal (command-line window) and issuing the command:

```
xmllint -noout pets.xml --schema pets.xsd
```

Nothing will appear on the screen, indicating that the *xml* file has been validated against the *xsd*.

2.10.4 Loss of freedom

It is clear that, by using a *dtd* or *xsd*, we give up our freedom to arbitrarily but sensibly define element tags, attributes, and nodes in an *xml* document. This is certainly disappointing, as lamented on previous occasions.

However, it must be emphasized that the use of a *dtd* or *xsd* is optional and relevant only at the stage where the *xml* data will be communicated or retrieved. In professional applications, *xml* authors are happy to conform with universal standards designed by others, so that their documents can be smoothly processed.

Exercise

2.10.1 *Dtd for a polynomial*

Write a *dtd* pertaining to the real, imaginary, or complex roots of an Nth-degree polynomial.

2.11 Xml namespaces

The finite-element and boundary-element methods are advanced numerical methods for solving differential equation in domains with arbitrary geometry.* The main advantage of the boundary-element method is that only the boundary of a given solution domain needs to be discretized into line elements in two dimensions or surface elements in three dimensions. The finite-element method is able to tackle a broader class of differential equations, albeit at a significantly elevated cost.

*Pozrikidis, C. (2008) *Numerical Computation in Science and Engineering*, Second Edition, Oxford University Press.

In both the finite-element and boundary-element method, geometrical elements with different shapes and sizes can be employed to accommodate the geometry of the solution domain. Examples include straight segments, circular arcs, triangles, and rectangles.

Finite-element and boundary-element grids

Assume that an *xml* document contains information on finite and boundary elements comprising corresponding grids used to solve the Laplace equation. It is desirable to use the name *element* in both cases, albeit in different contexts.

To achieve this, we introduce two *xml* namespaces (*xmlns*), one for the finite elements and the second for boundary elements. The content of the pertinent *xml* file named *elements.xml* is:

```
<?xml version="1.0" encoding="ISO-8859-1"?>
<laplace>

  <fem:grid xmlns:fem="femuri">
    <fem:element id="1" shape="triangular" nodes="3"/>
    <fem:element id="2" shape="rectangular" nodes="4"/>
  </fem:grid>

  <bem:grid xmlns:bem="bemuri">
    <bem:element id="1" shape="linear" nodes="2"/>
    <bem:element id="2" shape="circular" nodes="3"/>
  </bem:grid>

</laplace>
```

The name of the root element is `laplace`. Two uniform resource identifiers (*uri*) identified with two uniform resource names (*urn*), arbitrarily called *femuri* and *bemuri*, are used in this document. The *uris* are used to evaluate the *xmlns* attribute of the corresponding element, *xmlns:bem* and *xmlns:fem*. *Web* addresses called uniform resource locators (*url*) where information on each namespace is given are used in most applications as *uris*.

It is important to note that, if we had discarded the prefixes `fem:` and `bem:` in the *xml* file, we would no longer be able to distinguish between the finite- and boundary-element grids and identify the corresponding elements employed.

Alternatively, namespaces can be defined and evaluated as attributes of the root element of an *xml* document, as shown in the following file:

```
<?xml version="1.0" encoding="ISO-8859-1"?>
<laplace xmlns:fem="femuri" xmlns:bem="bemuri">

<fem:grid>
```

```
  <fem:element id="1" shape="triangular" nodes="3"/>
  <fem:element id="2" shape="rectangular" nodes="4"/>
</fem:grid>

<bem:grid>
  <bem:element id="1" shape="linear" nodes="2"/>
  <bem:element id="2" shape="circular" nodes="3"/>
</bem:grid>

</laplace>
```

The name of the root element is `laplace`.

In professional applications, namespaces are helpful when different parts of a code are written by different software engineering teams. Each team can be identified by its own namespace for credit or blame.

Default namespace

To simplify the notation, we may introduce a default namespace that lacks a prefix. Only one default namespace is allowed in an *xml* document. For example, a default finite-difference grid can be introduced by the following lines:

```
<grid xmlns="someuri">
  <xsize>16</xsize>
  <ysize>32</ysize>
</grid">
```

The finite-element namespace is the default namespace in the following *xml* document:

```
<?xml version="1.0" encoding="ISO-8859-1"?>
<laplace xmlns="femuri" xmlns:bem="bemuri">

  <grid>
    <element id="1" shape="triangular" nodes="3"/>
    <element id="2" shape="rectangular" nodes="4"/>
  </fem:grid>

  <bem:grid>
    <bem:element id="1" shape="linear" nodes="2"/>
    <bem:element id="2" shape="circular" nodes="3"/>
  </bem:grid>

</laplace>
```

Unless specified otherwise, all descendants of an element in the default namespace also fall in the default namespace.

Qualified names

An *xml* qualified name (*qname*) is an *xml* element name optionally preceded by a namespace, called the prefix, and a colon (`:`). An example is `bem:element`.

Exercise

2.11.1 *Namespaces*

Discuss a scientific or engineering application where the use of two namespaces is desirable.

2.12 Xml formatting of computer language instructions

A conditional block in a computer code has the following generic pseudocode structure:

```
If (something_is_true) then
  ...
End If
```

where the three lines represent instructions to be followed only if the statement `something_is_true` is true. A possible recasting of this block into *xml* compliant form is:

```
<if test="something_is_true">
  ...
</if>
```

The direct translation is possible because both the conditional block of the pseudocode and the `if` element of the *xml* document require closure. More generally, the syntax of any suitable computer language could be restated to comply with *xml* conventions.

As an example, we consider the following complete *fortran* program that defines and adds two numbers, and then prints their sum on the screen:

```
      program vasvas

      a = 5.87
      b = 6.01
      c = a+b
      print c

      stop
      end
```

Note that six mandatory blank spaces have been inserted at the beginning of each line. The stop statement refers to execution, and the end statement refers to compilation.

A possible equivalent code written in *xml* is:

```
<?xml version="1.0"?>

<fortran:program xmlns:fortran="primm">
  <fortran:variable type="real" name="a" value="5.87"/>
  <fortran:variable type="real" name="b" value="6.01"/>
  <fortran:variable type="real" name="c" value="a+b"/>
  <fortran:print format="any" name="c"/>
<fortran:program>
```

A program can be written that translates this *xml* document into *fortran* code, and *vice versa*. Although a complete set of semantics is conveyed by the *xml* document, the visual cluttering and the repetition of terms are distracting.

Xml programming language implementation

Programming languages that employ *xml* grammar and syntax are available. An example is the extensible stylesheet language (*xsl*) discussed in Chapters 3 and 4. A comprehensive list of other languages is available on the Internet.*

In using an *xml* compliant language, two *xml* files are necessary: an *xml* data file containing input, and an *xml* program file implementing language instructions. Each file has its own root element serving a different purpose. Conversely, an arbitrary *xml* file can contain either data or computer programming instructions.

Xsl

The typical structure of an *xsl* program file is:

```
<?xml version="1.0"?>
<xsl:stylesheet version="1.0"
  xmlns:xsl="http://www.w3.org/1999/XSL/Transform">
  ...
</xsl:stylesheet>
```

where the three dots indicate additional lines of code, as discussed in Chapters 3 and 4. The first line is an *xml* declaration. The second line introduces the root element named xsl:stylesheet and defines the *xsl* namespace. Other namespaces could be added to this line, if necessary. Since *xsl* statements adhere to *xml* standards, the first line of the code is not mandatory.

*http://en.wikipedia.org/wiki/List_of_XML_markup_languages

Scalable vector graphics (svg)

The scalable vector graphics (*svg*) suite defines a family of *xml* tags that describe geometrical elements representing two-dimensional vector graphics. *Svg* processors are embedded in *web* browsers and other graphics applications.

As an example, we consider the following *svg* file written in the *xml* format:

```
<?xml version="1.0" standalone="no"?>

<!DOCTYPE svg PUBLIC "-//W3C//DTD SVG 1.1//EN"
    "http://www.w3.org/Graphics/SVG/1.1/DTD/svg11.dtd">

<svg xmlns="http://www.w3.org/2000/svg" version="1.1">
  <circle cx="64" cy="64" r="32" stroke="red"
    stroke-width="2" fill="yellow" />
</svg>
```

Because an external *dtd* is employed, this *xml* file is not standalone. The name of the root element, qualified by two attributes, is `svg`. One *svg* element implementing a yellow disk enclosed by a red circle resides inside the root element. Opening the *xml* file with a *web* browser produces the following display:

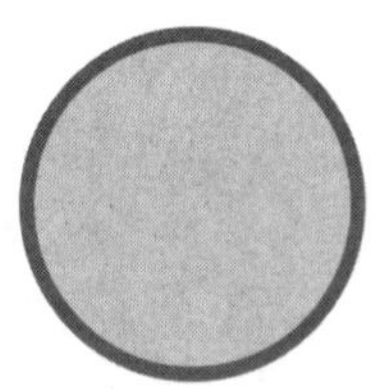

Svg and *mathml* have an element named *set*. Different namespaces must be used when *svg* and *mathml* are simultaneously employed.

tikzpicture

In typesetting this book, the previous display of the disk was programmed using the *tikzpicture latex* package, which also produces scalable vector graphics. The typesetting instructions in the *latex* document are:

```
\begin{centering}
    \begin{tikzpicture}
        \draw[fill=yellow] ellipse (32pt and 32pt);
        \draw[line width=1mm,color=red] circle (32pt);
    \end{tikzpicture}
\end{centering}
```

A variety of other *svg* graphics elements and corresponding *tikzpicture* elements are available.

Exercises

2.12.1 *Your computer language in xml*

Recast a scientific code in a language of your choice into *xml* format.

2.12.2 *Aragorn in xml*

Write a code of your choice in a fictitious computer language called *aragorn* according to *xml* conventions.

Xml data processing with xsl 3

Data contained in an *xml* document are typically manipulated by a computer program (*application*) written in a suitable computer language. The output is displayed on a screen, printed on paper, or stored in a file to be recalled at an opportune time. *Xml* data processing and manipulation are best illustrated with reference to the extensible stylesheet language (*xsl*) introduced in Section 1.5.2 and discussed in detail in this chapter. Scientific computing programmers should make a correspondence between an *xsl* processor and a computer language interpreter or compiler.

The basic features of the *xsl* language will be reviewed in this chapter and available procedures for data manipulation will be discussed and demonstrated by example. Applications of the *xsl* programming language in numerical computation will be discussed in Chapter 4.

Readers who are not particularly interested in the *xsl* programming language may skip Chapters 3 and 4 and proceed to Chapter 5 where methods of importing and exporting *xml* data from scientific code are outlined.

3.1 Xsl processors

To parse and process data contained in an *xml* file using an *xsl* processor, we write an *xsl* program file (code) containing instructions concerning data parsing and manipulation. Several processing options are available.

Xsltproc

An *xml* file can be transformed into another *xml* file according to a specified *xsl* file using the *xsltproc* application, available in most *linux* distributions and other operating systems. In fact, *xsltproc* is a command-line interface of the *libxml* parser and toolkit written by Daniel Veillard.* The application and its constituent libraries are written entirely in C. Installation on a *unix* system is straightforward using a software package manager.

*http://www.xmlsoft.org

Binary files implementing the *xsltproc* processor in Windows include executables and dynamic link libraries (*dll*).* The following executable files (*exe*) and dynamic link libraries (*dll*) are necessary for running *xsltproc* from a command-line window:

```
iconv.dll  libexslt.dll  libxslt.dll   xmlcatalog.exe  xsltproc.exe
iconv.exe  libxml2.dll   minigzip.exe  xmllint.exe     zlib1.dll
```

These files should be copied into directories (folders) that are included in the path of executables.

To process an *xml* data file named *myxmldata.xml*, subject to an accompanying *xsl* file referenced in the *xml* file by way of a processing instruction (PI), we issue the following statement in a terminal:

```
xsltproc myxmldata.xml -o mynewxmldata.xml
```

The output is recorded in a file named *mynewxmldata.xml*. If the optional argument

```
-o mynewxmldata.xml
```

is omitted, the output will be shown in the screen. If the *xml* file does not contain a processing instruction (PI), we type:

```
xsltproc myxslcode.xsl myxmldata.xml -o mynewxmldata.xml
```

Other optional features of *xsltproc* are available. To obtain further information on a *unix* system, we launch the manual pages by typing:

```
man xsltproc
```

On Windows, we issue the statement *xsltproc* in a command-line window.

Other xsl processors

Other *xsl* processors associated with advanced programming languages and frameworks are available, including *saxon*, *xalan*, and *xecres*.

Processing with a web browser

Xsl processors are embedded in standard *web* browsers. To process the *xml* data, we simply open the *xml* data file by selecting the *Open File* option in the *File* drop-down menu of a *web* browser. The name of the accompanying *xsl* file must be declared in the *xml* document through a processing instruction (PI). If the output of the *xsl* processor is *html* or other related code familiar to the browser, the processed formatted output will be shown in the browser window.

*http://www.zlatkovic.com/libxml.en.html

Exercise

3.1.1 *Summary*

Summarize the usage and features of an *xsl* processor of your choice.

3.2 The main program

Consider the following *xml* data file named *greet.xml* containing only three lines:

```
<?xml version="1.0" encoding="ISO-8859-1"?>
<?xml-stylesheet type="text/xsl" href="greet.xsl"?>
<hello/>
```

The first line declares the specific version of *xml* employed and the character encoding type, as discussed in Section 2.4. The second line is a processing instruction (PI) declaring the use of an *xsl* stylesheet (program). The name of the file where the stylesheet resides, *greet.xsl*, is specified as the hypertext reference attribute (*href*). The first two lines are standard in any *xml* document to be processed by an *xsl* file.

In our example, the main body of the *xml* document contains an obligatory root element named `hello` that happens to be self-closing. Normally, data and other nested tags implementing children elements will reside inside the root element, and element attributes will be present.

3.2.1 Xsl code

The accompanying *greet.xsl* file referenced in the processing instruction stated in the second line of the *xml* file contains the following lines:

```
<xsl:stylesheet version="1.0"
      xmlns:xsl="http://www.w3.org/1999/XSL/Transform">
<xsl:template match="hello">
<html>
    <b>Hello and welcome</b>
</html>
</xsl:template>
</xsl:stylesheet>
```

The first line continues onto the second line. The first and last statements mark the beginning and the end of the *xsl* code. These statements are mandatory in any *xsl* code.

Xsl employs xml syntax

It is remarkable that the *xsl* file is written according to *xml* grammar and syntax, as discussed in Chapter 2. In particular, the first statement:

```
<xsl:stylesheet version="1.0"
      xmlns:xsl="http://www.w3.org/1999/XSL/Transform">
```

defines the root element of the *xsl* file, regarded as an *xml* file, and specifies the *xsl* version number as an attribute. A dedicated *xsl* namespace is defined by the attribute *xmlns:xsl*, and a *web* address (*url*) is provided as a *uri* where further information can be obtained, as discussed in Section 2.11.

Xsl programming elements

Since am *xsl* file is also an *xml* document, it must be well-formed. All elements inside the root element of the *xsl* file must follow *xml* conventions and grammar. Thus, the grammar of *xsl* statements in an *xsl* file is similar that of *xml* elements in an *xml* data file, including the possible presence of element attributes. This similarity justifies referring to *xsl* statements as *xsl* elements. The terminology can be confusing, since a distinction must be made between *xml* data elements and *xsl* programming elements.

3.2.2 Root template

When we start processing the *greet.xml* file, control is passed to the *greet.xsl* code that launches a root template with a *match* attribute evaluated as the name of the root element of the *xml* file, `hello`, as specified in the second line of the *xsl* file under discussion:

```
<xsl:template match="hello">
```

The root template of the *xsl* code is a child of the root element of the *xsl* code. The *match* attribute specifies that the main program will employ data residing inside the root element of the *xml* file. Scientific computing programmers should immediately make the correspondence:

root template → *main program or main function*

For convenience, the root element of the *xml* data file can be denoted by a slash. Thus, the second line in the *xsl* file could be replaced by the line:

```
<xsl:template match="/">
```

The use of the slash to indicate a root (element or directory) is commonplace in *unix*.

A root template is mandatory

A root template referring to the root element of a processed *xml* file is mandatory in the corresponding *xsl* file. Similarly, a main function is mandatory in a C or C++ code and a main program is mandatory in a *fortran* code. One difference is that the root template in an *xsl* code must make reference to the root element of the *xml* file. In contrast, a C or C++ main function and a *fortran* main program are standalone. However, C, C++, and *fortran* must also declare at some stage the names of files containing input, if any.

Content of the root template

In our example, the *xsl* root template contains data and instructions that produce *html* code. The *html* interpreter is first launched by printing the `<html>` tag, which is implicitly equivalent to the more comprehensive tag:

```
<html xmlns="http://www.w3.org/1999/xhtml">
```

where the *html* namespace is defined as a default namespace. The words *Hello and welcome* are printed next in bold face, as instructed by the boldface *html* tag `<b>` and its closure `</b>`. Finally, the standard closing tag `</html>` is printed.

The root template and the stylesheet declaration close in the last two lines of the *xsl* file to render the *xsl* file well-formed.

3.2.3 Processing with the xsltproc processor

We will assume that the *xsl* program file resides in the same directory as the processed *xml* data file. This assumption will be implicit in our discussion in the remainder of this book. Issuing the following statement in a terminal (command-line window):

```
xsltproc greet.xml
```

displays on the screen the *html* code:

```
<html><b>Hello and welcome</b></html>
```

In this case, the output contains the entire content of the *xsl* file inside the root template. More generally, text contained in an *xsl* file will appear verbatim in the output when parsed by the *xsl* processor. Exceptions arise in the case of text implementing *xsl* elements (language commands) or logical blocks inside the root template.

We emphasize that, in this example, data are not transferred from the *xml* file to the *xsl* file. In a typical application, a wealth of data will be retrieved and processed by the *xsl* file.

3.2.4 Processing with a web browser

Opening the *greet.xml* file with a *web* browser produces the screen display:

Hello and welcome

Note that the text is printed in boldface, as indicated by the <b> (boldface) *html* tag and its closure </b>. Briefly, when we open the *greet.xml* file, the following actions take place:

1. The *xsl* processor embedded in the browser is informed that *xml* version 1.0 with ISO-8859-1 character encoding are employed.

2. The *xsl* processor is instructed to execute the root template implemented in the *greet.xsl* file, and thus parse the data contained inside the root element of the *xml* file.

3. Text and programming elements inside the root template of the *greet.xsl* file generate *html* statements.

4. The execution of the root template terminates.

5. The browser processes and displays the *html* code.

All but the last step are executed when the *xsltproc* processor is used instead of a *web* browser.

3.2.5 Comments

Comments can be inserted anywhere in an *xsl* file using standard *xml* delimiters, as discussed in Section 2.3. This is consistent with our earlier observation that an *xsl* program is also an *xml* document. For example, we may write:

```
<!-- Failure is simply the opportunity to begin again,
     this time more intelligently.
                                   Henry Ford -->
```

Multiple lines can be commented out using this convention for the purpose of debugging a code. It is rumored that professional code contains one error (bug) in every twenty lines.

Alternatively, comments can be inserted inside an *xsl* comment element in the format:

```
<xsl:comment>
  Please ignore me
</xsl:comment>
```

Further details on the *comment* programming element and other *xsl* elements are given in Appendix C.

3.2.6 File name conventions

The name of an *xsl* program file accompanying an *xml* data file can be arbitrary, bearing no relation to the name of the *xml* data file. However, the name of the *xsl* file must be referenced correctly in the *href* attribute of the processing instruction near the beginning the *xml* data file. Similarly, the name of the root element of the *xml* data file must be referenced correctly in the *match* attribute of the root template of the *xsl* file.

3.2.7 Inverting namespaces

In the example discussed in this section, we have used an explicit *xsl* namespace and an implicit default *html* namespace. The roles can be reversed by explicitly defining the *html* namespace and introducing an implicit default *xsl* namespace.

As an example, we consider the following *xml* document entitled *hiya.xml* containing a self-closing root element named `polite`:

```
<?xml version="1.0" encoding="ISO-8859-1"?>
<?xml-stylesheet type="text/xsl" href="hiya.xsl"?>
<polite/>
```

The accompanying *xsl* code residing in a file entitled *hiya.xsl*, referenced in the processing instruction stated in the second line, reads:

```
<stylesheet version="1.0"
            xmlns="http://www.w3.org/1999/XSL/Transform">
<template match="polite">

  <oaxaca:html xmlns:oaxaca="http://www.w3.org/1999/xhtml">
      <oaxaca:b>Hello and welcome</oaxaca:b>
  </oaxaca:html>

</template>
</stylesheet>
```

Note that the *xsl:* prefix is no longer used in front of the *xsl* elements. For example, *stylesheet* is used instead of *xsl:stylesheet* in the first line. The *html* namespace is named in the fifth line after a highly recommended Mexican cheese.

Opening the file *hiya.xml* with a *web* browser produces the screen display:

Hello and welcome

which is identical to that produced by the alternative code discussed earlier in this section.

Exercises

3.2.1 *Absence of a root template*

Discover what happens when the root template is omitted or the name of the root element is misprinted in the root template of an *xsl* file.

3.2.2 *Ascii art*

Insert the drawing of a cat in an *xsl* file using *ascii* art.

3.3 for-each loops

The procedure implemented in the *greet.xml* file and associated *greet.xsl* file discussed in Section 3.2 can be implemented in a different way. Consider the following *xml* document entitled *kalimera.xml*:

```
<?xml version="1.0" encoding="ISO-8859-1"?>
<?xml-stylesheet type="text/xsl" href="kalimera.xsl"?>
<willsayhello>

  <salutation/>

</willsayhello>
```

The name of the root element is `willsayhello`. The root element contains one self-closing tag named `salutation`, representing a solitary *xml* child. The accompanying *xsl* code contained in a file named *kalimera.xsl* reads:

```
<xsl:stylesheet version="1.0"
      xmlns:xsl="http://www.w3.org/1999/XSL/Transform">
<xsl:template match="willsayhello">
<html>

  <xsl:for-each select="salutation">
    <b>Good day</b>
  </xsl:for-each>

</html>
</xsl:template>
</xsl:stylesheet>
```

When we open the file *kalimera.xml* using a *web* browser, control is passed to the *kalimera.xsl* file that executes the root template based on the data contained inside the root element of the *xml* file. First, the *html* interpreter is launched. Second, the data residing inside the root element of the *xml* file are parsed from top to bottom in search of an instance of `salutation`. When an instance is found, the words *Good day* are printed in the output between *html* bold face tag

<b> and its closure </b>. Finally, the *html* interpreter is exited, the *html* code is processed by the *web* browser, and the outcome is displayed in the browser.

Opening the *kalimera.xml* file with a *web* browser produces the following display:

Good day

In summary, the `for-each` tag and its closure implement an *xsl* programming element that may enclose other *xsl* elements of the same or different namespace.

Event-driven action

Scientific programmers will recognize the structure:

```
<xsl:for-each select="salutation">
  ...
</xsl:for-each>
```

as a *Do* loop in *fortran* or a *for* loop in C, C++, or *Matlab*, where the three dots indicate additional lines of code. For example, in *fortran*, we write

```
Do i=1,10
  ...
End Do
```

where i is a dummy index running from 1 to 10 with default increment of 1.

In *xsl*, the range of repetition of an index in a *for-each* loop is unspecified and determined by the content of the *xml* data file. This peculiarity underlines the dominance of the data in the *xml* framework. Adding another line:

```
<salutation/>
```

before the last line of the *kalimera.xml* file, and opening the revised file with a *web* browser, prints in the window of the browser a duplicate message in bold face:

Good day Good day

The reason for the duplication is that two instances of salutation are found inside the root element of the *xml* file. If a third instance were present, the message would be repeated one more time.

We say that *xml*/*xsl* is an event-driven framework, where the events are recorded in the *xml* data file.

Nested loops

`for-each` loops referring to data trees can be nested. As an example, we consider the following *xml* file entitled *apresto.xml*:

```
<?xml version="1.0" encoding="ISO-8859-1"?>
<?xml-stylesheet type="text/xsl" href="apresto.xsl"?>
<greeting>

  <salutation>
    <thanks/>
  </salutation>

</greeting>
```

The name of the root element is `greeting`. The accompanying *apresto.xsl* file referenced in the second line reads:

```
<xsl:stylesheet version="1.0"
      xmlns:xsl="http://www.w3.org/1999/XSL/Transform">
<xsl:template match="greeting">
<html>

  <xsl:for-each select="salutation">
    hello
    <xsl:for-each select="thanks">
      there
    </xsl:for-each>
  </xsl:for-each>

</html>
</xsl:template>
</xsl:stylesheet>
```

When we open the file *apresto.xml* with a *web* browser, control is passed to the *apresto.xsl* file that executes the root template based on data residing inside the root element of the *xml* file. First, the *html* interpreter is launched. Second, the data inside the root element of the *xml* file are parsed from top to bottom in search of an instance of *salutation*. If an instance is found, the word *hello* is printed in the standard font. If an instance of `thanks` is found inside an instance of `salutation`, the word *there* is also printed in the standard font.

Opening the *apresto.xml* file with a *web* browser produces the expected screen display:

```
hello there
```

Note that, once we find ourselves inside the first *for-each* loop, we make local element name selections. Thus, replacing inside the second *for-each* loop the line:

```
<xsl:for-each select="thanks">
```

with the revised line:

```
<xsl:for-each select="salutation/thanks">
```

would be wrong. The reason is that the revision implies the presence of an instance of salutation inside an instance of salutation (not that this would be wrong).

Scientific programmers will note the similarity between the *for-each* nested structure and the following generic nested structure of a scientific code:

```
for i=1:10
    for j=1:14
    ...
    end
end
```

where the three dots represent additional lines of code. To underscore the main difference, we emphasize again that the range of the repetition index in a *for-each* loop of an *xsl* code is determined by the content of the processed *xml* data file in an event-driven framework.

Exercises

3.3.1 *Two loops*

Write a code that employs two sensible sequential `for-each` loops.

3.3.2 *Good night*

Modify the code *kalimera.xsl* so that *and good night* is printed after *Good day.*

3.4 Extracting data with value-of

To extract data from an *xml* file, we use the *value-of* programming element. The data can be numbers, characters strings, words, sentences, or longer text.

As an example, we consider the following *xml* data contained in a file named *diaduit.xml*:

```
<?xml version="1.0" encoding="ISO-8859-1"?>
<?xml-stylesheet type="text/xsl" href="diaduit.xsl"?>
```

```
<greeting>

  <salutation>
    Good Day
  </salutation>

</greeting>
```

The name of the root element is `greeting`. An *xml* node named `salutation` containing pure text resides inside the root element. The accompanying program file *diaduit.xsl* referenced in the second line reads:

```
<xsl:stylesheet version="1.0"
      xmlns:xsl="http://www.w3.org/1999/XSL/Transform">
<xsl:template match="greeting">
<html>

  <xsl:for-each select="salutation">
    <xsl:value-of select="."/>
  </xsl:for-each>

</html>
</xsl:template>
</xsl:stylesheet>
```

When we open the file *diaduit.xml* with a *web* browser, control is passed to the *diaduit.xsl* file that executes the root template based on data residing inside the *greeting* root element. First, the *html* interpreter is launched. Second, the material residing inside the root element of the *xml* file is parsed from top to bottom in search of an instance of `salutation`. When an instance is found, the content of `<salutation>`, that is, the text enclosed between the `<salutation>` tag and its closure, `</salutation>`, indicated by the dot in the line:

```
<xsl:value-of select="."/>
```

is printed in the output. In this case, the text reads *Good Day.* Opening the *diaduit.xml* file with a *web* browser produces the screen display:

```
Good Day
```

This important example illustrates that specific *xml* data can be extracted from an *xml* file in a way that is unfamiliar to scientific computing programmers. The self-closing tag `value-of` implements an *xsl* element that carries information by way of the *select* attribute. We recall that, in *xsl*, the dot indicates the content of the current *xml* node. In the *unix* operating system, one dot denotes the current directory, and two dots denote the parental directory.

Use of entities

As another example, we consider an *xml* file entitled *cars.xml*, containing the following declarations and data:

```
<?xml version="1.0" encoding="ISO-8859-1"?>
<?xml-stylesheet type="text/xsl" href="cars.xsl"?>

<!DOCTYPE inventory [
  <!ENTITY found "Found a " >
  <!ENTITY built " built in the year " >
]>

<inventory>

  <car>
    <make>&found; Wartburg</make>
    <year>&built; 1966</year>
  </car>
  <car>
    <make>&found; Scania</make>
    <year>&built; 1969</year>
  </car>

</inventory>
```

The name of the root element is `inventory`. An internal *dtd* defining two entities is implemented at the beginning of the file following the *xsl* processing instruction stated in the second line.

The accompanying *cars.xsl* file referenced in the second line reads:

```
<xsl:stylesheet version="1.0"
      xmlns:xsl="http://www.w3.org/1999/XSL/Transform">
<xsl:template match="inventory">
<html>

<xsl:for-each select="car">
  <xsl:value-of select="make"/>
  <xsl:value-of select="year"/>
  <br/>
</xsl:for-each>

</html>
</xsl:template>
</xsl:stylesheet>
```

Note that, once we find ourselves inside the *for-each* loop, we make local element name selections. Opening the *cars.xml* file with a *web* browser prints on the screen:

```
Found a Wartburg built in the year 1966
Found a Scania built in the year 1969
```

We observe that the entities referenced in the *xml* data file are printed as defined in the *dtd.*

Exercise

3.4.1 *Pressure and humidity*

Write an *xml* file that contains the temperature and humidity of the atmosphere at a certain location each day over a period of a week. Then write a companion *xsl* file that reports these data in a grammatically correct sentence.

3.5 Repeated parsing

An *xml* file can be parsed in the same session repeatedly multiple times, each time performing a different task. Different data can be extracted each time the *xml* document is parsed.

As an example, we consider data contained in the following file entitled *dogsandcats.xml*:

```
<?xml version="1.0" encoding="ISO-8859-1"?>
<?xml-stylesheet type="text/xsl" href="dogsandcats.xsl"?>
<kennel>

  <dog>Dalmatian</dog>
  <cat>Main coon</cat>
  <dog>Golden retriever</dog>

</kennel>
```

The name of the root element, `kennel`, implies that these adorable creatures temporarily live in a kennel.

The accompanying *dogsandcats.xsl* file referenced in the second line reads:

```
<xsl:stylesheet version="1.0"
      xmlns:xsl="http://www.w3.org/1999/XSL/Transform">
<xsl:template match="kennel">
<html>

  <br/>DOGS:<br/>
  <xsl:for-each select="dog">
    <xsl:value-of select="."/> <br/>
```

```
  </xsl:for-each>
  <br/>

  <br/>CATS:<br/>
  <xsl:for-each select="cat">
    <p><u><xsl:value-of select="."/></u></p> <br/>
  </xsl:for-each>

</html>
</xsl:template>
</xsl:stylesheet>
```

When we open the *dogsandcats.xml* file with a *web* browser, control is passed to the *dogsandcats.xsl* file that executes the root template using data contained inside the `kennel` root element of the *xml* file. The following actions take place:

1. The *html* interpreter is launched.
2. The text "DOGS:" is printed.
3. The *xsl* processor parses the material inside the root element of the *xml* file from top to bottom in search of instances of *dog*. When an instance is found, the text enclosed between the `<dog>` tag and its closure is printed.
4. The text "CATS:" is printed.
5. The *xsl* processor parses once again the material inside the root element of the *xml* file from top to bottom in search of instances of *cat*. When an instance is found, the text enclosed between the `<cat>` tag and its closure `</cat>` defining another *xml* node, is printed.
6. The generated *html* code is processed by the browser and printed on the screen.

Opening the *animals.xml* file with a *web* browser produces the screen display:

This example emphasizes further that multiple sets of specific data can be extracted from a data file. In the present case, the extraction is done sequentially by multiple parsings, rather than concurrently in a single parsing. In a scientific application, an *xml* document could contain information on velocity and temperature. We may first extract and visualize the data on the velocity, and then extract and visualize the data on the temperature.

Dogs and cats can been printed in the output in order of appearance in the *xml* data file by using templates, as discussed in Section 3.10.

Exercise

3.5.1 *Velocity and temperature*

Write an *xml* document that contains data on the velocity and temperature at the nodes of a finite-difference grid, and an accompanying *xsl* code that retrieves the temperature and then the velocity.

3.6 Extracting element attributes

Anything enclosed between an *xml* tag and its closure is an object that can be assigned an attribute, such as an identification number (id).

Consider the following data contained in the following file entitled *farm.xml*:

```
<?xml version="1.0" encoding="ISO-8859-1"?>
<?xml-stylesheet type="text/xsl" href="farm.xsl"?>
<animals>

  <cow id="1">Molly</cow>
  <horse id="1">Zebediah</horse>
  <cow id="2">Gateway</cow>
  <horse id="2">Biscuit</horse>

</animals>
```

The name of the root element is `animals`. The accompanying *farm.xsl* file referenced in the second line reads:

```
<xsl:stylesheet version="1.0"
      xmlns:xsl="http://www.w3.org/1999/XSL/Transform">
<xsl:template match="animals">
<html>

  COWS:
  <p/>
  <xsl:for-each select="cow">
    <xsl:value-of select="@id"/>:
    <xsl:value-of select="."/>
    <br/>
  </xsl:for-each>
  <p/>
  HORSES:
  <p/>
  <xsl:for-each select="horse">
```

```
    <xsl:value-of select="@id"/>:
    <xsl:value-of select="."/>
    <br/>
  </xsl:for-each>

</html>
</xsl:template>
</xsl:stylesheet>
```

Prepending the character (@) to an attribute name extracts the value of the attribute, which can be a number or character string. Opening the *farm.xml* file with a *web* browser produces the screen display:

```
COWS:

1: Molly
2: Gateway

HORSES:

1: Zebediah
2: Biscuit
```

In scientific applications, ids can be assigned to finite or boundary elements in finite- or boundary-element codes, as illustrated in the following example.

Use of namespaces

Information on a finite- and a boundary-element grid is contained in the following file entitled *laplace.xml* under two namespaces, as discussed in Section 2.11:

```
<?xml version="1.0" encoding="ISO-8859-1"?>
<?xml-stylesheet type="text/xsl" href="laplace.xsl"?>
<laplace xmlns:fem="femuri" xmlns:bem="bemuri">

  <fem:grid>
    <fem:element id="1" shape="triangular" nodes="3"/>
    <fem:element id="2" shape="rectangular" nodes="4"/>
  </fem:grid>

  <bem:grid>
    <bem:element id="1" shape="linear" nodes="2"/>
    <bem:element id="2" shape="circular" nodes="3"/>
  </bem:grid>

</laplace>
```

The name of the root element is `laplace`. Two namespaces are introduced by way of the *xmlns:fem* and *xmlns:bem* attributes of the root element. These namespaces are evaluated as uniform resource identifiers *uris*, arbitrarily named *femuri* and *bemuri*. The accompanying *laplace.xsl* file referenced in the second line reads:

```
<xsl:stylesheet version="1.0"
      xmlns:xsl="http://www.w3.org/1999/XSL/Transform"
                xmlns:fem="femuri"
                xmlns:bem="bemuri">
<xsl:template match="laplace">
<html>

<!-- boundary elements -->

<xsl:for-each select="bem:grid">
  <xsl:for-each select="bem:element">
    One <xsl:value-of select="@shape"/> boundary element with
    <xsl:value-of select="@nodes"/> nodes found
    (element index:  <xsl:value-of select ="@id"/>) <br/>
  </xsl:for-each>
</xsl:for-each>

<br/>

<!-- finite elements -->

<xsl:for-each select="fem:grid">
  <xsl:for-each select="fem:element">
    One <xsl:value-of select="@shape"/> finite element with
    <xsl:value-of select="@nodes"/> nodes found
    (element index:  <xsl:value-of select ="@id"/>) <br/>
  </xsl:for-each>
</xsl:for-each>

<!--- done -->

</html>
</xsl:template>
</xsl:stylesheet>
```

Note that the two namespace *uris*, named *femuri* and *bemuri*, are reproduced faithfully in the root element of the *xsl* file. Opening the *laplace.xml* file with a *web* browser produces the screen display:

```
One linear boundary element with 2 nodes found (element index: 1)
One circular boundary element with 3 nodes found (element index: 2)

One triangular finite element with 3 nodes found (element index: 1)
One rectangular finite element with 4 nodes found (element index: 2)
```

This example illustrates that namespaces play the role of surnames of *xml* elements in an *xsl* code for the purpose of unambiguous identification.

Exercise

3.6.1 *Animal age and nickname*

Modify the *farm.xsl* code given in text to print the age and nickname of each animal. Both should be defined in the *xml* file as element attributes.

3.7 Conditional blocks

Conditional blocks are commonplace in *xsl* programming. Consider the data contained in the following file entitled *positive.xml*:

```
<?xml version="1.0" encoding="ISO-8859-1"?>
<?xml-stylesheet type="text/xsl" href="farm.xsl"?>
<surplus>

  <number>12</number>
  <number>-4</number>
  <number>23</number>

</surplus>
```

The name of the root element is `surplus`. The accompanying *positive.xsl* file reads:

```
<xsl:stylesheet version="1.0"
      xmlns:xsl="http://www.w3.org/1999/XSL/Transform">
<xsl:template match="surplus">
<html>

  <xsl:for-each select="number">
    <xsl:if test=".&gt;0">
      <xsl:value-of select="."/>
    </xsl:if>
  </xsl:for-each>

</html>
</xsl:template>
</xsl:stylesheet>
```

The expression

```
. &gt; 0
```

should be read as *current value is greater than zero.* We note that the *greater*

xsl alias	Implied symbol	Meaning
=	$\leftarrow$	replace left by right
=	$=$	equal to
!=	$\neq$	not equal to
<	$<$	less than
>	$>$	greater than
®	®	registered
&	&	ampersand
"	"	double quotation
'	'	single quotation
©	©	copyright

TABLE 3.1 *Xsl* aliases of special symbols are listed in the left column. Note the semicolon (;) at the end of the last seven aliases.

than mathematical operator (>) is encoded as `>`. Other standard encodings are shown in Table 3.1. Further mathematical operators are discussed in Chapter 4.

In our example, the code inside the *if* element is executed only if the expression tested is true. With reference to *xml* grammar, *test* is an attribute of the *if* element evaluated by the programmer.

Opening the file *positive.xml* with a *web* browser produces the screen display:

```
12 23
```

Only positive numbers are printed.

In summary, the `if` tag and its closure implement an *xsl* element enclosing other elements of the same or different namespace.

Composite and negation tests

Composite tests can be made using logical `and` and logical `or` operators. Negation tests can be performed using the negation operator, `not()`. In our example, we may perform the test:

```
<xsl:if test=".&gt;0 and .&lt;20">
```

which assesses whether a current value, indicated by the dot, is greater than 0 and less than 20. Opening the file *positive.xml* with a *web* browser produces the screen display:

```
12
```

Only numbers in the range 0–20 are printed.

The tested expression

```
not(.&lt;0)
```

establishes whether the current value, indicated by the dot, is not negative, which means that it is zero or positive (≥ 0). The equivalent *fortran* statement is `a.ge.0`, where a is a variable carrying a value of interest. Similar statements in other programming languages can be written using the operants shown in Table 4.1.

Exercise

3.7.1 *Constellations*

Write an *xml* document that contains the names of ten constellations and provides the number of stars in each constellation as an element attribute. Then write an *xsl* file that extracts and prints the names of constellations with less than five stars.

3.8 Choose, when, and otherwise

`If/else` blocks are commonplace in scientific computer programming. In *xsl*, these logical constructs are implemented as `choose`, `when`, and `otherwise` logical blocks.

As an example, we consider the following *budget.xml* file containing one negative and two positive numbers:

```
<?xml version="1.0" encoding="ISO-8859-1"?>
<?xml-stylesheet type="text/xsl" href="budget.xsl"?>
<bliss>

  <number>-2.3</number>
  <number>11.8</number>
  <number>2.2</number>

</bliss>
```

The name of the root element is `bliss`. The associated *budget.xsl* file reads:

```
<xsl:stylesheet version="1.0"
      xmlns:xsl="http://www.w3.org/1999/XSL/Transform">
<xsl:template match="bliss">
```

```
<html>

<xsl:for-each select="number">

  <xsl:choose>

    <xsl:when test=". &lt; 0">
      ( <xsl:value-of select="."/> )
    </xsl:when>

    <xsl:when test=". &gt; 10">
      [ <xsl:value-of select="."/> ]
    </xsl:when>

    <xsl:otherwise>
      <xsl:value-of select="."/>
    </xsl:otherwise>

  </xsl:choose>
</xsl:for-each>

</html>
</xsl:template>
</xsl:stylesheet>
```

Opening the *budget.xml* file with a *web* browser produces the screen display:

```
(-2.3) [11.8] 2.2
```

Negative numbers are enclosed by parentheses, positive numbers higher than 10 are enclosed by square brackets, and all other numbers are printed plainly.

Exercise

3.8.1 *Integrals*

Write an *xml* document that contains a list of ten named functions, such as the exponential function, and indicates with an attribute whose value can be 0 or 1 whether the indefinite integral of each function is known in closed form. Then write an *xsl* program that prints the name of the function next to the message *known indefinite integral* or *unknown indefinite integral.*

3.9 Variables and parameters

In scientific programming, we introduce numerical variables, define their type (integer, single precision, or double precision), and then evaluate the variables by literal or data assignment. Similar procedures are available in the *xml* framework.

Unfortunately, once an *xsl* variable has been introduced and evaluated, it cannot be reevaluated in terms of its current value. The inability to update a variable is the main reason that *xsl* is classified as a functional language, as opposed to a procedural language, such as *fortran*, as discussed in Section 4.2.

Local variables

As an example, we consider the results of an experiment recorded in an *xml* document contained in a file entitled *variables.xml*:

```
<?xml version="1.0" encoding="ISO-8859-1"?>
<?xml-stylesheet type="text/xsl" href="variables.xsl"?>
<experiment_A>

  <response id="1">13.3</response>
  <response id="2">15.4</response>

</experiment_A>
```

The name of the root element is `experiment_A`. The accompanying file entitled *variables.xsl* referenced in the second line reads:

```
<xsl:stylesheet version="1.0"
      xmlns:xsl="http://www.w3.org/1999/XSL/Transform">
<xsl:template match="experiment_A">
<html>

  <xsl:for-each select="response">
    <xsl:value-of select="."/> is the same as

<!-- by variable:  -->

    <xsl:variable name="a">
      <xsl:value-of select="."/>
    </xsl:variable>

    <xsl:value-of select="$a" /><br/>

  </xsl:for-each>

</html>
</xsl:template>
</xsl:stylesheet>
```

Each response is retrieved and printed using the *value-of* programming element.

For demonstration, a variable `a` is introduced and evaluated as the numerical content of the current response. The value of the variable, denoted by `$a`, is then printed.

Opening the *variables.xml* file with a *web* browser produces the screen display:

```
13.3 is the same as 13.3
15.4 is the same as 15.4
```

The *xsl* processor recognizes that the variable `a` holds a numerical value rather than a character string. Conversion of a numerical character string into a numerical value is possible, as discussed in Section 3.10.

We can introduce, name, and evaluate by literal a variable by stating, for example,

```
<xsl:variable name="pi" select="3.14159265358"/>
```

The variable may then be printed using the statement:

```
<xsl:value-of select="$pi"/>
```

Definition and evaluation by literal is useful when a variable is, in fact, a constant.

Global variables

A global variable can be defined immediately after the *stylesheet* declaration and before the root template statement in an *xsl* document. A global variable is accessible to all templates (function or subroutines), as discussed in Sections 3.10 and 3.11.

Parameters

Parameters are variables used by templates representing user-defined functions and subroutines. The declaration, usage, and passing of parameters are discussed in Sections 3.10 and 3.11.

Exercise

3.9.1 *Response id*

Modify the *xsl* code given in the text so that the response id is printed sensibly in the output.

3.10 Templates

We have seen that a root template (main program) associated with the root element of the *xml* file is mandatory in an *xsl* file. Other lower-level templates can be defined and executed (applied). Templates are similar to functions and subroutines employed in scientific computer programming, in that they receive input and produce output. An important feature of *xsl* templates is that they are granted full implicit access to all data contained in a processed *xml* document. *Xsl* employs matched templates and named templates to achieve complementary and overlapping goals.

3.10.1 Matched templates

Consider the data contained in the following *xml* file entitled *autos.xml*:

```
<?xml version="1.0" encoding="ISO-8859-1"?>
<?xml-stylesheet type="text/xsl" href="autos.xsl"?>
<inventory>

  <car>Ford</car>
  <truck>Chevrolet</truck>
  <car>GM</car>

</inventory>
```

The name of the root element is `inventory`. The accompanying *autos.xsl* file referenced in the second line prints car and trucks:

```
<xsl:stylesheet version="1.0"
  xmlns:xsl="http://www.w3.org/1999/XSL/Transform">
<xsl:template match="inventory">
<html>

  <xsl:for-each select="car">
    Auto:  <xsl:value-of select="."/> <br/>
  </xsl:for-each>

  <xsl:for-each select="truck">
    Camion:  <xsl:value-of select="."/> <br/>
  </xsl:for-each>

</html>
</xsl:stylesheet>
```

Opening the *autos.xml* file with a *web* browser produces the expected screen display:

```
Auto: Ford
Auto: GM
Camion: Chevrolet
```

Note that, since the *xml* data are parsed sequentially twice, first cars and then trucks are displayed.

In order of appearance

Now we want to inspect each element inside the root element of the *xml* file in order of appearance and take appropriate action. For this purpose, we parse the *xml* data from top to bottom and subject each element to a series of tests by applying matched templates. The procedure is implemented in the following *autos.xsl* file defining templates for cars and trucks:

```
<xsl:stylesheet version="1.0"
      xmlns:xsl="http://www.w3.org/1999/XSL/Transform">
<xsl:template match="inventory">
<html>

  <xsl:apply-templates/>

</html>
</xsl:template>

<!-- user-defined matched template:  car -->

  <xsl:template match="car">
    Car:  <xsl:value-of select="."/><br/>
  </xsl:template>

<!-- user-defined matched template:  truck -->

  <xsl:template match="truck">
    Truck:  <xsl:value-of select="."/><br/>
  </xsl:template>

<!-- end of user-defined matched templates -->

</xsl:stylesheet>
```

Opening the file *autos.xml* with a *web* browser produces the display:

```
Auto: Ford
Truck: Chevrolet
Auto: GM
```

Note that cars and trucks are displayed in order of appearance in the *xml* file, as desired.

When we open the *autos.xml* file with a *web* browser, control is passed to the *autos.xsl* file that executes the root template using data contained inside the `inventory` root element of the *xml* file. All templates defined in the *xsl* file are then applied to each element inside the root element of the *xml* file in order of appearance, as instructed by the self-closing *xsl* element (declaration):

```
xsl:apply-templates
```

If a template match is found, that is, if the name of the *xml* node matches the value of the match attribute stated in the template, the action specified inside the template is taken.

Scientific programmers will recognize that a matched template is a function or subroutine that receives input from a specific element of an *xml* file. However, the idea of subjecting data to different templates for the purpose of finding a match is novel to scientific programmers who are used to running templates (functions or subroutines) with specified data, as opposed to querying data by different methods.

Absence of templates

What happens if no templates are implemented in an *xsl* file? Absent a default template, as discussed in the next section, the *xsl* processor will print the content of the parsed elements. In our example, opening the *autos.xml* file with a *web* browser will produce the display:

```
Ford Chevrolet GM
```

Default template

An unimplemented matched template reverts to a default template, if present, whose general statement is:

```
<xsl:template match="*">
  ...
</xsl:template>
```

where the three lines indicate additional lines of code. In computer programming, an asterisk (*) is known as a wildcard. As an example, we replace the two templates in the *autos.xsl* file with a single default template defined as:

```
<xsl:template match="*">
  Vehicle:  <xsl:value-of select="."/> <br/>
</xsl:template>
```

Opening the *autos.xml* file with a *web* browser produces the display:

```
Vehicle: Ford
Vehicle: Chevrolet
Vehicle: GM
```

In practice, a default template is used to avoid repetitive code in cases where the same action is taken on multiple elements of an *xml* document.

Priority

Priority must be assigned when two or a higher number of templates match the same *xml* element according to the following syntax:

```
<xsl:template match="xml_element_name" priority="index">
  ...
</xsl:template>
```

where the three dots indicate additional lines of code, and ***index*** is a real number in the range $[-9.0, 9.0]$ with default value 0. The ***priority*** attribute is optional.

Matched templates with parameters

Matched templates with parameters can be regarded as functions or subroutines with arguments that convey information that is complementary to that contained in the processed *xml* file.

As an example, the following *xml* document entitled *dimless.xml* contains the names of dimensionless numbers in fluid mechanics and heat transport:

```
<?xml version="1.0" encoding="ISO-8859-1"?>
<?xml-stylesheet type="text/xsl" href="dimless.xsl"?>
<dimensionless>

  <fluid_mech>Reynolds</fluid_mech>
  <heat_trans>Peclet</heat_trans>

</dimensionless>
```

The name of the root element is `dimensionless`. The companion *dimless.xsl* file reference in the second line reads:

```
<xsl:stylesheet version="1.0"
      xmlns:xsl="http://www.w3.org/1999/XSL/Transform">
<xsl:template match="dimensionless">
<html>

<xsl:apply-templates>
  <xsl:with-param name="before" select="'The ' "/>
  <xsl:with-param name="after" select="' number appears in ' "/>
</xsl:apply-templates>

</html>
</xsl:template>
```

```
<!-- user-defined matched template:  fluid_mech -->

<xsl:template match="fluid_mech">
  <xsl:param name="before"/>
  <xsl:param name="after"/>
    <xsl:value-of select="$before"/> <xsl:value-of select="."/>
    <xsl:value-of select="$after"/> fluid mechanics<br/>
</xsl:template>

<!-- user-defined matched template:  heat_trans -->

<xsl:template match="heat_trans">
  <xsl:param name="before"/>
  <xsl:param name="after"/>
    <xsl:value-of select="$before"/> <xsl:value-of select="."/>
    <xsl:value-of select="$after"/> heat transfer<br/>
</xsl:template>

<!-- end of user-defined matched templates -->

</xsl:stylesheet>
```

Opening the *dimnum.xsl* file with a *web* browser produces the display:

```
The Reynolds number appears in fluid mechanics
The Peclet number appears in heat transfer
```

In this example, matched templates with two parameters named `before` and `after` have been applied. To access the values of these parameters, we prepended the dollar sign (`$`) to their name inside the templates.

Applying specific matched templates

The personnel of an auto shop is listed in the following *shop.xml* file:

```
<?xml version="1.0" encoding="ISO-8859-1"?>
<?xml-stylesheet type="text/xsl" href="shop.xsl"?>
<personnel>

  <mechanic>Jack Wilson</mechanic>
  <cashier>Harry Smith</cashier>
  <mechanic>Rebecca Smith</mechanic>
  <mechanic>Samantha Stewart</mechanic>
  <supervisor>Mr.  Bean</supervisor>

</personnel>
```

The name of the root element is `personnel`. To identify and print only the mechanics in the shop, we use the following *shop.xsl* file:

```
<xsl:stylesheet version="1.0"
      xmlns:xsl="http://www.w3.org/1999/XSL/Transform">
<xsl:template match="personnel">
<html>

  <xsl:apply-templates select="mechanic"/>

</html>
</xsl:template>

<!-- user-defined matched template:  mechanic -->

  <xsl:template match="mechanic">
    Mechanic <xsl:value-of select="."/> <br/>
  </xsl:template>

<!-- end of user-defined template -->

</xsl:stylesheet>
```

We see that a selected template alone is applied inside the root template of the *xsl* file using the statement:

```
<xsl:apply-templates select="mechanic"/>
```

Opening the *shop.xml* file with a *web* browser produces the screen display:

```
Mechanic Jack Wilson
Mechanic Rebecca Smith
Mechanic Samantha Stewart
```

In the algorithm implemented in the *xsl* code, each *xml* element is subjected to a single selected matched template. If a match is found, action is taken. If a match is not found, action is not taken.

As an experiment, we insert after the line:

```
<xsl:apply-templates select="mechanic"/>
```

inside the root template of the *xsl* file the new line:

```
<xsl:apply-templates select="cashier"/>
```

Opening the *shop.xml* file produces the display:

```
Mechanic Jack Wilson
Mechanic Rebecca Smith
Mechanic Samantha Stewart
Harry Smith
```

This experiment confirms that, in the case of an unimplemented template, the browser simply prints the content of the matched element.

3.10.2 Named templates

The matched templates discussed in Section 3.10.1 refer to selected *xml* element nodes. Named templates can be defined without explicit reference to *xml* element nodes by the statements:

```
<xsl:template name="somename">
  <xsl:param name="argument1"/>
  <xsl:param name="argument2"/>
  ...
  <xsl:param name="argument7"/>
    ......
</xsl:template>
```

where the three dots indicate additional parameters and the six dots indicate additional lines of code implementing the template. The name of the template, `somename`, bears no relationship to the name of the *xml* elements parsed by the template. Scientific programmers recognize named templates as functions or subroutines accompanying a program or main function.

To call a named template, we state in the *xsl* file:

```
<xsl:call-template name="somename">
  <xsl:with-param name="argument1" select="value1"/>
  <xsl:with-param name="argument2" select="value2"/>
    ...
  <xsl:with-param name="argument7" select="value7"/>
</xsl:call-template>
```

where the three dots indicate additional parameters conveying data in addition to those recorded in the parsed *xml* document.

Mechanics

As an example, we consider information on mechanics working in an auto repair shop contained in the following *xml* file named *mechanics.xml*:

```
<?xml version="1.0" encoding="UTF-8"?>
<?xml-stylesheet type="text/xsl" href="mechanics.xsl"?>
<mechs>
```

```
  <mechanic certified="YES">Ashley Moore</mechanic>
  <mechanic certified="YES">George Fine</mechanic>

</mechs>
```

The name of the root element is `mechs`. The data are processed by the following *mechanics.xsl* file referenced in the second line:

```
<xsl:stylesheet version="1.0"
      xmlns:xsl="http://www.w3.org/1999/XSL/Transform">
<xsl:template match="mechs">
<html>

  <xsl:for-each select="mechanic">
    <xsl:call-template name="details"/>
  </xsl:for-each>

</html>
</xsl:template>

<!-- user-defined named template:  details -->

<xsl:template name="details">
    Mechanic <xsl:value-of select="."/>,
    Certified:  <xsl:value-of select="@certified"/><br/>
</xsl:template>

<!-- end of user-defined template -->

</xsl:stylesheet>
```

Opening the *mechanics.xml* file with a *web* browser prints on the screen:

```
Mechanic Ashley Moore, Certified: YES
Mechanic George Fine, Certified: YES
```

In this example, the name and certified attribute of each mechanic are printed by way of a named template. Consistent with our earlier remarks, the template has access to all information pertinent to each mechanic.

Data accessibility

We have demonstrated that data contained in an *xml* document is fully accessible to all templates. As a further demonstration, we consider information on engineers in the research and development division of a company, contained in the following *xml* file entitled *eng.xml*:

```
<?xml version="1.0" encoding="UTF-8"?>
<?xml-stylesheet type="text/xsl" href="eng.xsl"?>
<smolikas>

  <engineer>Terry Gibbons</engineer>
  <engineer>Terri Manolis</engineer>

</smolikas>
```

The name of the root element is `smolikas`. The accompanying *eng.xsl* file referenced in the second line reads:

```
<xsl:stylesheet version="1.0"
      xmlns:xsl="http://www.w3.org/1999/XSL/Transform">
<xsl:template match="smolikas">
<html>

  <xsl:call-template name="print_name"/>

</html>
</xsl:template>

<!-- user-defined template:  print_name -->

<xsl:template name="print_name">
  <xsl:for-each select="engineer">
    Engineer <xsl:value-of select="."/><br/>
  </xsl:for-each>
</xsl:template>

<!-- end of user-defined template -->

</xsl:stylesheet>
```

Note a `for-each` loop inside the template. Opening the *eng.xml* file with a *web* browser prints on the screen:

```
Engineer Terry Gibbons
Engineer Terri Manolis
```

The names of the engineers are extracted and printed inside the named template.

Return of a scalar

A named template receives input by way of parameters and *xml* elements, but is unable to return output. A scalar value can be returned only if a template is called from within a variable definition. In contrast, standard scientific programming languages allow functions and subroutines to return diverse output.

To illustrate the return of a scalar, we refer to the *eng.xsl* code and replace the programming element:

```
<xsl:call-template name="print_name"/>
```

stated in the sixth line of the *xsl* file, with the lines:

```
<xsl:variable name="saved_names">
  <xsl:call-template name="print_name"/>
</xsl:variable>

<xsl:value-of select="$saved_names"/><br/>
```

Opening the *eng.xml* file with a *web* browser produces on the screen the display:

```
Engineer Terry Gibbons Engineer Terri Manolis
```

The output of the template is returned and printed as the value of a variable, in this case a character string.

3.10.3 Matched and named templates

The general statement of a matched and named template is:

```
<xsl:template match="xml_node_name" name="somename"
                    priority="index" mode ="modename">
  <xsl:with-param name="somename" select="some_expression"/>
  ...
  <xsl:with-param name="lastname" select="last_expression"/>
    ......
</xsl:template>
```

We have seen that all matched templates are executed using the statement:

```
<xsl:apply-templates>
```

and one chosen matched template can be executed using the statement:

```
<xsl:apply-templates select="nodename">
```

A named template is executed using the statement:

```
<xsl:call-template name="templatename">
```

A matched and named template can be executed either way.

Cooks

As an example, we consider the following file entitled *cook.xml* containing the names of cooks in a tavern:

```
<?xml version="1.0" encoding="UTF-8"?>
<?xml-stylesheet type="text/xsl" href="cook.xsl"?>
<arathorn>

  <cook>Sam Tam</cook>
  <cook>Tim Smith</cook>

</arathorn>
```

The root element is named after the tavern owner, `arathorn`. The accompanying *cook.xsl* file referenced in the processing instruction stated in the second line reads:

```
<xsl:stylesheet version="1.0"
     xmlns:xsl="http://www.w3.org/1999/XSL/Transform">
<xsl:template match="arathorn">
<html>

  <xsl:for-each select="cook">
    <xsl:call-template name="print_name"/>
  </xsl:for-each>
  <br/>

  <xsl:apply-templates select="cook"/>

</html>
</xsl:template>

<!-- user defined-template:  print_name -->

<xsl:template match="cook" name="print_name">
  Cook <xsl:value-of select="."/> works here<br/>
</xsl:template>

<!-- end of user-defined template -->

</xsl:stylesheet>
```

Opening the *cook.xsl* file with a *web* browser prints on the screen a duplicate human resources listing:

```
Cook Sam Tam works here
Cook Tim Smith works here
```

```
Cook Sam Tam works here
Cook Tim Smith works here
```

Exercises

3.10.1 *Matched templates with empty default template*

Discuss the output of a procedure where all matched templates are applied, one specific matched template is implemented, and an idle default template is present. Generalize the discussion to address the case where a number of specific matched templates are implemented.

3.10.2 *Named and matched template*

Write a code that employs in a meaningful fashion a matched and named template.

3.11 Splitting the code

Templates can be conveniently placed in separate files and imported into a main *xsl* stylesheet referenced in an *xml* data file.

As an example, we consider information on authors of graduate textbooks in fluid mechanics contained in the following *xml* file entitled *authors.xml*:

```
<?xml version="1.0" encoding="UTF-8"?>
<?xml-stylesheet type="text/xsl" href="authors.xsl"?>
<tymfristos>

  <author>G.K. Batchelor</author>
  <author>C. Pozrikidis</author>

</tymfristos>
```

The name of the root element is `tymfristos`. The accompanying *authors.xsl* code referenced in the second line reads:

```
<xsl:stylesheet version="1.0"
      xmlns:xsl="http://www.w3.org/1999/XSL/Transform">
<xsl:import href="authors_tmpl.xsl"/>
<xsl:template match="tymfristos">
<html>

  <xsl:call-template name="print_name"/>

</html>
</xsl:template>
</xsl:stylesheet>
```

A stylesheet entitled *authors_tmpl* is imported in the third line by the programming element:

```
<xsl:import href="authors_tmpl.xsl"/>
```

The content of the imported stylesheet is:

```
<xsl:stylesheet version="1.0"
      xmlns:xsl="http://www.w3.org/1999/XSL/Transform">

<!-- user-defined template:  print_name -->

<xsl:template name="print_name">
  <xsl:for-each select="author">
  Author <xsl:value-of select="."/><br/>
  </xsl:for-each>
</xsl:template>

<!-- end of user-defined template -->

</xsl:stylesheet>
```

Note that an imported stylesheet lacks a root template.

Opening the *authors.xml* file with a *web* browser prints on the screen:

```
Author G.K. Batchelor
Author C. Pozrikidis
```

The author names are extracted and printed inside the named template `print_name` implemented in the imported template.

Exercise

3.11.1 *Code splitting*

Split a code of your choice into two files containing, respectively, the root template and user-defined templates.

3.12 Summary of xsl elements and functions

We have discussed several *xslt* programming elements, such as the *for-each* element. Scientific programmers will recognize these elements as computer language statements, directives, and constructs. A comprehensive summary of available *xslt* elements is given in Appendix C. Further information can be found in resources available on the Internet.

Orthogonal polynomials

As an application, we consider the following file entitled *orthopoly.xml* containing the names of inventors of orthogonal polynomials:

```
<?xml version="1.0" encoding="ISO-8859-1"?>
<?xml-stylesheet type="text/xsl" href="orthopoly.xsl"?>
<orthopolis>

  <poly>Legendre</poly>
  <poly>Jacobi</poly>
  <poly>Lagrange</poly>
  <poly>Hermite</poly>
  <poly>Chebyshev</poly>

</orthopolis>
```

The name of the root element is `orthopolis`. The accompanying *xsl* file entitled *orthopoly.xsl* reads:

```
<xsl:stylesheet version="1.0"
      xmlns:xsl="http://www.w3.org/1999/XSL/Transform">
<xsl:template match="orthopolis">
<html>

  <xsl:for-each select="poly">
    <xsl:sort select="." order="descending"/>
    <xsl:value-of select="."/>
    <xsl:text> </xsl:text>
  </xsl:for-each>

</html>
</xsl:template>
</xsl:stylesheet>
```

Opening the *orthopoly.xml* file with a *web* browser prints on the screen:

```
Legendre Lagrange Jacobi Hermite Chebyshev
```

Space between words was inserted using the `text` element:

```
<xsl:text> </xsl:text>
```

We see that the polynomials are listed in descending alphabetical order in the output, thanks to the `sort` element. This example illustrates that sorted elements are printed only after all *xslt* instructions have been processed and rearranged to ensure proper ordering at the end.

Linear equations

As a second example, we consider the following file entitled *linear.xml* containing the names of methods for solving systems of linear equations:

```
<?xml version="1.0" encoding="ISO-8859-1"?>
<?xml-stylesheet type="text/xsl" href="linear.xsl"?>
<karabouzouklis>

    <method>Gauss elimination</method>
    <method>LU decomposition</method>
    <method>Jacobi iterations</method>

</karabouzouklis>
```

The name of the root element is **karabouzouklis**. The accompanying file *linear.xsl* referenced in the second line reads:

```
<xsl:stylesheet version="1.0"
      xmlns:xsl="http://www.w3.org/1999/XSL/Transform">
<xsl:output method="xml" indent="yes"/>
<xsl:template match="karabouzouklis">

<xsl:copy>

  <xsl:text>&#10;&#10;</xsl:text>
  <xsl:comment> Methods for solving linear systems </xsl:comment>
  <xsl:text>&#10;&#10;</xsl:text>

  <xsl:for-each select="method">
    <xsl:copy-of select="."/>
    <xsl:text>&#10;</xsl:text>
  </xsl:for-each>

  <xsl:element name="method">
    <xsl:attribute name="recommended">
      <xsl:text>yes</xsl:text>
    </xsl:attribute>
    SOR iterations
  </xsl:element>

</xsl:copy>

</xsl:template>
</xsl:stylesheet>
```

Issuing the command:

```
xsltproc linear.xml
```

in a terminal (command-line window) produces the display:

```
<?xml version="1.0"?>
<karabouzouklis>

<!-- Methods for solving linear systems -->

<method>Gauss elimination</method>
<method>LU decomposition</method>
<method>Jacobi iterations</method>
<method recommended="yes">
SOR iterations
</method>

</karabouzouklis>
```

We see that a new *xml* file with altered elements has been produced. Several *xslt* programming elements are used in the *xsl* code. The name of the root element was preserved thanks to the `copy` programming element. A line break was forced by printing the corresponding character,

```
<xsl:text>&#10;</xsl:text>
```

3.12.1 *Core functions*

Internal (core) *xslt* and *xpath* functions are available for evaluating the attributes of *xslt* programming elements, as discussed in Appendix D.

Counting engineers

As an example, the following *xml* file entitled *factory.xml* contains the names of chemical and mechanical engineers working in a factory:

```
<?xml version="1.0" encoding="UTF-8"?>
<?xml-stylesheet type="text/xsl" href="factory.xsl"?>
<personnel>

  <chemeng>Kate Wilson</chemeng>
  <mecheng>Kleanthis Malamataris</mecheng>
  <chemeng>Nicole Dubois</chemeng>

</personnel>
```

The name of the root element is `personnel`. The companion *factory.xsl* file referenced in the second line reads:

```
<xsl:stylesheet version="1.0"
    xmlns:xsl="http://www.w3.org/1999/XSL/Transform">
<xsl:template match="personnel">
```

```
<html>

  <xsl:value-of select="count(chemeng)"/>
    chemical engineers work in this factory

</html>
</xsl:template>
</xsl:stylesheet>
```

Opening the *factory.xml* file with a *web* browser prints on the screen:

```
2 chemical engineers work in this factory
```

We see that both chemical engineers have been counted, thanks to the `count` function.

Testing for numbers

To assess whether the content of the current *xml* node indicated by a dot is a number, we perform the following test:

```
<xsl:if test="string(number(.))='NaN'">
  ...
</xsl:if>
```

where the three dots represent additional lines of code. The *number* and *string* functions are employed in this example.

Exercises

3.12.1 *Sorting*

Write an *xsl* code that sorts alphabetically the names of your favorite ethnic dishes contained in an *xml* file.

3.12.2 *Counting*

Write an *xsl* code that counts the names of your favorite ethnic dishes contained in an *xml* file.

3.13 Passive processing and cascading stylesheets (css)

To appreciate the versatility of the extensible stylesheet (*xsl*), we contrast it with the cascading stylesheet (*css*) used in *html* programming. Consider the following file entitled *memo.xml* encapsulating a text message:

```
<?xml version="1.0"?>
<?xml-stylesheet type="text/css" href="memo.css"?>
```

```
<memo>

  <from>FROM: J. Smith</from>
  <to>TO: W. Williams</to>
  <reference>RE: Adaptive self-evaluation</reference>
  <emailtext>Please explain the concept of metacognition.</emailtext>

</memo>
```

The name of the root element is `memo`. The second line is a processing instruction (PI) referring to the cascading stylesheet (*css*) file *memo.css* whose content is:

```
memo
{ background-color:yellow; font-size:10pt; }
from
{ Display:block; color:olive; margin-left:20pt; }
to
{ Display:block; color:#FF00FF; margin-left:20pt; }
reference
{ Display:block; color:blue; margin-left:20pt; }
emailtext
{ Display:block; color:blue; margin-top:20pt; margin-left:20pt; }
```

The *css* file specifies the *html* properties of each element introduced in the *xml* file: `memo, from, to, reference, emailtext.` For example, the statement `Display:block;` specifies that a line break will be generated before and after an element is displayed. Opening the *memo.xsl* with a *web* browser produces the display:

```
FROM: J. Smith
TO: W. Williams
RE: Adaptive self-evaluation

Please explain the concept of metacognition.
```

where different lines are printed with different color in yellow background.

We see that a *css* file can be used to conveniently format the content of an *xml* file for viewing in a *web* browser. However, information processing is passive, in that the data cannot be altered, repeated, or given individual attention. The *xml*/*css* pair has no advantages over the *html*/*css* pair.

Exercise

3.13.1 *Bibliographical information*

Write a *css* file that prints a book reference based on data contained in an *xml* file using a display format of your choice.

4 Computing with *xml/xsl*

In Chapters 1 and 2 we discussed the motivation behind *xml* formatting and introduced the basic *xml* grammar. In Chapter 3 we explained how an *xml* document can be processed according to a companion *xsl* program to produce desired output that can be stored in a file or displayed in a *web* browser. In science and engineering applications, we are primarily interested in arithmetic data manipulation and to a lesser extent on formatted display. In this chapter, we discuss numerical computation in the *xml/xsl* framework.

The case studies discussed in this chapter will confirm the assertion that, intended to be a *web* programming and data manipulation language, *xsl* is hardly appropriate for scientific computation and should be used only for elementary calculations. Nevertheless, the lack of numerical facilities in the *xsl* processor is welcome as a challenge for exercising creativity and a motivation for sharpening programming skills.

4.1 Elementary operations

Elementary mathematical operations in an *xsl* code are implemented by the operants:

```
+     -     *     div
```

for addition, subtraction, multiplication, and division, respectively.

It is important to keep in mind that expressions on either side of the subtraction operator (`-`) must be separated by white space. Otherwise, the minus sign will be misinterpreted as a dash connecting characters, and an error message will be issued by the *xsl* processor. For example, `4 - 5` will return `-1`, but `4-5` will not be recognized as an arithmetic manipulation.

The modulo (*mod*) operator returns the numerator of the fractional remainder in division. For example, since `13/4 = 3 + 1/4`, the operation `13 mod 4` returns `1`.

Relational and logical operands of interest in scientific computing are summarized in Table 4.1 in six programming languages, including *xsl*. The equal

<table>
<tr><th>Operation</th><th>Matlab</th><th>Fortran</th><th>C++</th><th>Perl strings</th><th>Perl numbers</th><th>Xsl</th></tr>
<tr><td>add</td><td>+</td><td>+</td><td>+</td><td>.</td><td>+</td><td>+</td></tr>
<tr><td>subtract</td><td>-</td><td>-</td><td>-</td><td></td><td>-</td><td>-</td></tr>
<tr><td>multiply</td><td>*</td><td>*</td><td>*</td><td></td><td>*</td><td>*</td></tr>
<tr><td>divide</td><td>/</td><td>/</td><td>/</td><td></td><td>/</td><td>div</td></tr>
<tr><td>exponentiation</td><td>^</td><td>**</td><td>pow</td><td></td><td>**</td><td></td></tr>
<tr><td>modulo</td><td>mod</td><td>mod</td><td>%</td><td></td><td>%</td><td>mod</td></tr>
<tr><td>replace</td><td>=</td><td>=</td><td>=</td><td>=</td><td>=</td><td>=</td></tr>
<tr><td>equal</td><td>==</td><td>=</td><td>==</td><td>eq</td><td>==</td><td>=</td></tr>
<tr><td>not equal</td><td>~=</td><td>.ne.</td><td>!=</td><td>ne</td><td>!=</td><td>!=</td></tr>
<tr><td>less</td><td><</td><td>.lt.</td><td><</td><td>lt</td><td><</td><td><</td></tr>
<tr><td>less or equal</td><td><=</td><td>.le.</td><td><=</td><td>le</td><td><=</td><td>not(>)</td></tr>
<tr><td>greater</td><td>></td><td>.gt.</td><td>></td><td>gt</td><td>></td><td>></td></tr>
<tr><td>greater or equal</td><td>>=</td><td>.ge.</td><td>>=</td><td>ge</td><td>>=</td><td>not(<)</td></tr>
<tr><td>and</td><td>&</td><td>.and.</td><td>&&</td><td>and</td><td>&&</td><td>and</td></tr>
<tr><td>or</td><td>|</td><td>.or.</td><td>||</td><td>or</td><td>||</td><td>or</td></tr>
</table>

TABLE 4.1 Relational and logical operands in *Matlab*, *fortran*, C and C++, *perl*, and *xsl*. Note that the *Matlab* and C++ columns are nearly identical.

sign (=) normally implements a left-by-right replace. Thus, the statement a=b requests replacing the value of a with the value of b. In C, C++, *Matlab*, and *perl*, a distinction is made between the replace operator (=) and the equal operator (==). In *fortran* and *xsl*, the same symbol is employed.

As an example, we add two numbers contained in the following *xml* file named *addition.xml*:

```
<?xml version="1.0" encoding="ISO-8859-1"?>
<?xml-stylesheet type="text/xsl" href="addition.xsl"?>
<algebra>

  <number1>34.09</number1>
  <number2>-17.12</number2>

</algebra>
```

The name of the root element is algebra. The companion *addition.xsl* file referenced in the second line selects and adds the two numbers and prints their sum:

```
<xsl:stylesheet version="1.0"
      xmlns:xsl="http://www.w3.org/1999/XSL/Transform">
<xsl:template match="algebra">
<html>

  <xsl:value-of select="number1"/> + <xsl:value-of select="number2"/>
      = <xsl:value-of select="number1 + number2"/>

</html>
</xsl:template>
</xsl:stylesheet>
```

Opening the *addition.xml* file with a *web* browser produces the screen display:

34.09 + -17.12 = 16.970

The *xsl* processor embedded in the browser recognizes that the contents of the elements `number1` and `number2` are numbers, not character strings, and performs sensible numerical addition instead of string concatenation. If we try to add a number and a character string, clearly defying the rules of algebra, we will obtain `NaN` (Not a Number) in the output.

The *xsl* code contained in the *addition.xsl* file appears similar to a *fortran* or C++ code in that it fetches data from the *xml* file and performs a calculation using the *xml* processor embedded in the *web* browser as a computational engine.

4.1.1 Using variables

With reference to our last example, precisely the same output can be obtained by assigning each number to be added to a variable, and then adding and printing the values of the two variables according to the following code:

```
<xsl:stylesheet version="1.0"
      xmlns:xsl="http://www.w3.org/1999/XSL/Transform">
<xsl:template match="algebra">
<html>

  <xsl:variable name="n1">
    <xsl:value-of select="number1"/>
  </xsl:variable>

  <xsl:variable name="n2">
    <xsl:value-of select="number2"/>
  </xsl:variable>

  <xsl:value-of select="$n1"/> + <xsl:value-of select="$n2"/>
     = <xsl:value-of select="$n1 + $n2"/>
```

```
</html>
</xsl:template>
</xsl:stylesheet>
```

Recall that the value of a variable is extracted by prepending the dollar sign ($) to the variable name.

Once we have evaluated numerical variables, we can combine them according to the rules of algebra using elementary operators. The practice of introducing variables and then performing arithmetic manipulations is familiar to scientific programmers.

A calculator

In another example, we consider the following *xml* file entitled *calculator.xml* defining two numbers and then requesting addition and multiplication:

```
<?xml version="1.0" encoding="ISO-8859-1"?>
<?xml-stylesheet type="text/xsl" href="calculator.xsl"?>
<calc>

  <number1>13.09</number1>
  <number2>07.12</number2>
  <add/>
  <multiply/>

</calc>
```

The name of the root element is `calc`. In the companion *calculator.xsl* file listed below, the two numbers are used to evaluate two variables, a and b, the variables are added and multiplied, and the result of each operation is recorded in the output:

```
<xsl:stylesheet version="1.0"
      xmlns:xsl="http://www.w3.org/1999/XSL/Transform">
<xsl:template match="calc">
<html>

  <xsl:variable name="a">
    <xsl:value-of select="number1"/>
  </xsl:variable>

  <xsl:variable name="b">
    <xsl:value-of select="number2"/>
  </xsl:variable>

  <xsl:for-each select="add">
```

```
    <xsl:value-of select="$a"/>+<xsl:value-of select="$b"/>
    =<xsl:value-of select="$a + $b"/><br/>
  </xsl:for-each>

  <xsl:for-each select="multiply">
    <xsl:value-of select="$a"/>*<xsl:value-of select="$b"/>
    =<xsl:value-of select="$a * $b"/><br/>
  </xsl:for-each>

</html>
</xsl:template>
</xsl:stylesheet>
```

For clarity, but not by necessity, empty spaces were inserted on either side of selected + and * operators. Opening the *calculator.xml* file with a *web* browser produces the screen display:

```
13.09+07.12 = 20.21
13.09*07.12 = 93.2008
```

In this example, *xml* elements are used to convey numerical data (numbers) or launch arithmetic operations (add and multiply).

4.1.2 Internal mathematical functions

A limited number of mathematical functions are included in the *xslt* 1.0 and *xpath* 1.0 processors, as discussed in Appendix D. The function *count* can be used to count similar nodes, and the function *sum* can be used to sum numerical nodes. Combining these two functions allows us to compute averages of numbers provided as *xml* nodes, as illustrated in the following example.

Computing an average (count and sum)

The following file entitled *avg.xml* contains a few velocity and temperature measurements recorded as responses:

```
<?xml version="1.0" encoding="ISO-8859-1"?>
<?xml-stylesheet type="text/xsl" href="exp.xsl"?>
<aristotelis>

<response>
  <velocity>12.0</velocity>
  <temperature>65.0</temperature>
</response>

<response>
  <velocity>13.1</velocity>
  <temperature>69.0</temperature>
```

```
</response>

<response>
  <velocity>14.9</velocity>
  <temperature>68.0</temperature>
</response>

</aristotelis>
```

The name of the root element is `aristotelis`. The data are processed according to an *xsl* code contained in the following file entitled *avg.xsl*:

```
<xsl:stylesheet version="1.0"
      xmlns:xsl="http://www.w3.org/1999/XSL/Transform">
<xsl:template match="aristotelis">
<html>

<xsl:variable name="nresp">
  <xsl:value-of select="count(response)"/>
</xsl:variable>

<xsl:value-of select="$nresp"/> responses received
<br/>
<xsl:value-of select="sum(response/velocity) div $nresp"/>
is the average velocity

<br/>
<xsl:value-of select="sum(response/temperature) div $nresp"/>
is the average temperature
<br/>

</html>
</xsl:template>
</xsl:stylesheet>
```

A variable named *nresp* is introduced and evaluated using the *count* function. The sums of the velocity and temperature values are computed by the *sum* function with an argument that is set equal to the corresponding node set. Arithmetic operations are then combined to produce the desired output. Opening the *avg.xml* file with a *web* browser produces the display:

```
3 responses received
13.333333333333334 is the average velocity
67.33333333333333 is the average temperature
```

4.1.3 Formatting numbers in the output

The numerical display can be formatted properly using available *xsl* functions. Several methods of formatting numbers in the output are available based on the following and other string functions:*

`round` `ceiling` `floor` `format-number`

As an example, we consider two numbers recorded in the following file entitled *format.xml*:

```
<?xml version="1.0" encoding="ISO-8859-1"?>
<?xml-stylesheet type="text/xsl" href="format.xsl"?>
<karaghiozis>

  <number1>5.0</number1>
  <number2>3.0</number2>

</karaghiozis>
```

The name of the root element is **karaghiozis**. The numbers are processed according to a code contained in the following *format.xsl* file:

```
<xsl:stylesheet version="1.0"
      xmlns:xsl="http://www.w3.org/1999/XSL/Transform">
<xsl:template match="karaghiozis">
<html>

<xsl:variable name="ratio">
  <xsl:value-of select="number1 div number2"/>
</xsl:variable>

<xsl:value-of select="$ratio"/> <br/>
<xsl:value-of select="round($ratio)"/> <br/>
<xsl:value-of select="ceiling(100 * $ratio) div 100"/> <br/>
<xsl:value-of select="floor(100 * $ratio) div 100"/> <br/>
<xsl:value-of select="format-number($ratio, '##.#####')"/>

</html>
</xsl:template>
</xsl:stylesheet>
```

Opening the *format.xml* file with a *web* browser generates the following display:

*Mangano, S. (2005) *XSLT Cookbook.* Second Edition, O'Reilly.

```
1.6666666666666667
2
1.67
1.66
1.66667
```

The *format-number* function provides us with the most direct method of selecting a desired numerical format and precision.

4.1.4 Maximum and minimum

The following *xml* file named *minmax.xml* contains several numbers:

```
<?xml version="1.0" encoding="ISO-8859-1"?>
<?xml-stylesheet type="text/xsl" href="minmax.xsl"?>
<trahanas>

  <arithmos>2.0</arithmos>
  <arithmos>5.0</arithmos>
  <arithmos>-3.0</arithmos>
  <arithmos>0.3</arithmos>

</trahanas>
```

The name of the root element is `trahanas`. The accompanying *minmax.xsl* file identifies and prints the maximum and minimum based on the *sort* element of the *xslt* processor and the *position* core function:

```
<xsl:stylesheet version="1.0"
      xmlns:xsl="http://www.w3.org/1999/XSL/Transform">
<xsl:template match="trahanas">
<html>

The maximum is:

<xsl:for-each select="arithmos">
  <xsl:sort data-type="number" select="." order="descending"/>
  <xsl:if test = "position() = 1">
    <xsl:value-of select="."/>
  </xsl:if>
</xsl:for-each>
<br/>

The minimum is:

<xsl:for-each select="arithmos">
  <xsl:sort data-type="number" select="." order="ascending"/>
  <xsl:if test = "position() = 1">
    <xsl:value-of select="."/>
```

```
  </xsl:if>
</xsl:for-each>

</html>
</xsl:template>
</xsl:stylesheet>
```

Opening the *minmax.xml* file with a *web* browser generates the display:

```
The maximum is: 5.0
The minimum is: -3.0
```

Other methods of extracting the maximum and minimum are available.*

4.1.5 Counting our blessings

In the absence of an extensive mathematical library and available data structures, we are left with the interesting yet intriguing task of building templates that implement low- and mid-level mathematical operations using elementary procedures. Implementations rely heavily on the use of named templates, as discussed in the remainder of this chapter.

Exercises

4.1.1 *Calculator*

Modify the *calculator.xsl* stylesheet so that the results of subtraction and division are also shown in the output.

4.1.2 *Computing a sum*

Investigate whether it is possible to add an arbitrary list of numbers recorded in an *xml* document without using the `sum` core function.

4.2 Templates are user-defined functions and subroutines

Scientific computing programmers appreciate the usefulness and versatility of user-defined functions. *Xsl* user-defined functions are matched or named templates, and their arguments are called parameters, as discussed in Section 3.10. We recall that templates have full implicit access to the data contained in a processed *xml* file. Named templates with parameters are most useful in scientific computing.

Xsl is a functional (non-imperative) programming language, in that the recursive use of templates (functions or subroutines) plays a prominent role. In an imperative programming language, we may issue the instruction:

*Mangano, S. (2005) *XSLT Cookbook.* Second Edition, O'Reilly.

```
x = x + 3
```

which implements a left-by-right replace, that is, it requests replacing the current value of x by itself plus 3, thus updating a given state. This operation is not possible in *xsl*, and we must rely on the judicious use of templates to perform relative simple tasks, such as computing a sum.

4.2.1 Absolute value of a number

In the first application, we are interested in printing the absolute value of each number recorded in the following file entitled *abs.xml*:

```
<?xml version="1.0" encoding="ISO-8859-1"?>
<?xml-stylesheet type="text/xsl" href="norm.xsl"?>
<maintanos>

  <number>10.0</number>
  <number>-2.0</number>

</maintanos>
```

The name of the root element is `maintanos`. The data will be processed according to instructions contained in the following *norm.xsl* file referenced in the second line:

```
<xsl:stylesheet version="1.0"
      xmlns:xsl="http://www.w3.org/1999/XSL/Transform">

<!-- compute the absolute value of a real number -->

<xsl:template match="maintanos">
<html>

<xsl:for-each select="number">

  The absolute value of <xsl:value-of select="."/> is
  <xsl:call-template name="absolute">
    <xsl:with-param name="x" select="."/>
  </xsl:call-template>
  <br/>

</xsl:for-each>

</html>
</xsl:template>

<!-- user-defined template:  absolute -->
```

```
<xsl:template name="absolute">
    <xsl:param name="x"/>

    <xsl:choose>

      <xsl:when test="$x &gt; 0">
        <xsl:value-of select="$x"/>
      </xsl:when>

      <xsl:otherwise>
        <xsl:value-of select="-$x"/>
      </xsl:otherwise>

    </xsl:choose>

</xsl:template>

<!-- end of user-defined template -->

</xsl:stylesheet>
```

A named template called `absolute` with a single parameter named `x` is employed in the *xsl* code. The *value-of* programming element is used in the template to print the value of the variable `x` if `x` is positive, and the negative of the value of `x` otherwise.

Opening the *abs.xml* file with a *web* browser produces the display:

```
The absolute value of 10.0 is 10.0
The absolute value of -2.0 is 2
```

This example illustrates that a parameter is automatically treated as a variable whose value is extracted by prepending the dollar sign ($) to the name of the parameter.

4.2.2 Binary representation of a fractional number

To find the binary representation of the fractional number 0.28125, we compute the following sequential products:

$$0.28125 \times 2 = \mathbf{0}.5625, \qquad 0.5625 \times 2 = \mathbf{1}.125, \qquad 0.125 \times 2 = \mathbf{0}.250,$$
$$0.25 \times 2 = \mathbf{0}.500, \qquad 0.5 \times 2 = \mathbf{1}.000.$$

The calculations terminate when the fractional part has become zero. Taking the integer bold-faced figures in forward order, we obtain the binary representation

$$0.28125 = (0.01001)_2 \,.$$

This means that

$$0.28125 = 0 \times 2^{-1} + 1 \times 2^{-2} + 0 \times 2^{-3} + 0 \times 2^{-4} + 1 \times 2^{-5}.$$

Five binary digits (bits) are necessary to represent this fractional number on the right side of the binary point. An infinite number of bits is necessary to represent an arbitrary fractional number.

The following file entitled *bif.xml* contains several fractional numbers:

```
<?xml version="1.0" encoding="ISO-8859-1"?>
<?xml-stylesheet type="text/xsl" href="bif.xsl"?>
<prespa>

  <number>0.250</number>
  <number>0.500</number>
  <number>0.345</number>
  <number>0.999</number>

</prespa>
```

The name of the root element is `prespa`. The data will be processed according to instructions contained in the following *bif.xsl* file:

```
<xsl:stylesheet version="1.0"
      xmlns:xsl="http://www.w3.org/1999/XSL/Transform">
<xsl:template match="prespa">
<html>

<!-- binary representation of a number < 1 to n bits -->

<xsl:for-each select="number">
  The binary representation of <xsl:value-of select="."/> is:
    0.<xsl:call-template name="reduce">
        <xsl:with-param name="x" select="2 *(. - floor(.))"/>
        <xsl:with-param name="n" select="10"/>
    </xsl:call-template> <br/>
</xsl:for-each>

</html>
</xsl:template>

<!-- user-defined template:  reduce -->

<xsl:template name="reduce">
  <xsl:param name="x"/>
  <xsl:param name="n"/>

  <xsl:if test="$n &gt; 0">
```

```
      <xsl:value-of select="floor($x)"/>
      <xsl:call-template name="reduce">
        <xsl:with-param name="x" select="2 * ($x - floor($x))"/>
        <xsl:with-param name="n" select="$n - 1"/>
      </xsl:call-template>
    </xsl:if>

  </xsl:template>

  <!-- end of user-defined template -->

  </xsl:stylesheet>
```

The code implements the decimal-to-binary conversion algorithm discussed earlier in this section using the *floor* internal function. In the *xsl* implementation, a user-defined template named `reduce` is applied recursively with two parameters, `x` and `n`. Each time a binary digit is found and printed, the parameter `n` is reduced by one unit. The computations terminate when `n` has become zero.

Opening the *binary.xml* file with a *web* browser produces the display:

```
The binary representation of 0.250 is: 0.0100000000
The binary representation of 0.500 is: 0.1000000000
The binary representation of 0.345 is: 0.0101100001
The binary representation of 0.999 is: 0.1111111110
```

The results in the first two lines confirm that, when a number is multiplied by 2, the binary point is shifted to the right by one place.

Like C and C++, but unlike *fortran*, the *xsl* language supports recursive function calling. This means that a template can call itself an unspecified number of times to produce an intended nested sequence. In fact, recursive function calling is an essential feature of *xsl*. Without it, we would be extremely limited in our ability to implement even simple numerical algorithms.

4.2.3 Binary representation of any number

We have noted that, each time a number is multiplied by 2, the binary point is shifted to the left by one place. Conversely, each time a number is multiplied by 1/2 (divided by 2), the binary point is shifted to the right by one place.

These observations provide us with a practical method of deducing the binary representation of an arbitrary number based on the algorithm discussed in Section 4.2.2 for a fractional number: (a) a given number is repeatedly divided by 2 until it becomes less than 1, (b) the binary representation of the transformed number is found, and (c) the binary point is shifted to the right by the number of divisions carried out.

As an example, we consider the following *xml* data contained in a file entitled *binary.xml* containing two numbers:

```
<?xml version="1.0" encoding="ISO-8859-1"?>
<?xml-stylesheet type="text/xsl" href="binary.xsl"?>
<anargiros>

  <number>414.341</number>
  <number>0.191</number>

</anargiros>
```

The name of the root element is **anargiros**. The accompanying *binary.xsl* file referenced in the second line implementing the decimal-to-binary conversion algorithm is listed below:

```
<xsl:stylesheet version="1.0"
     xmlns:xsl="http://www.w3.org/1999/XSL/Transform">

<!-- compute the binary representation of any number with n bits -->

<xsl:template match="anargiros">
<html>

<xsl:for-each select="number">
  The binary representation of <xsl:value-of select="."/> is:
    <xsl:call-template name="normalize">
      <xsl:with-param name="x" select="."/>
      <xsl:with-param name="p" select="0"/>
    </xsl:call-template>
    <br/>
</xsl:for-each>

</html>
</xsl:template>

<!-- user-defined template:  normalize -->

<xsl:template name="normalize">
  <xsl:param name="x"/>
  <xsl:param name="p"/>

  <xsl:choose>

  <xsl:when test="$x &gt; 1.0">
    <xsl:call-template name="normalize">
      <xsl:with-param name="x" select="0.5 * $x"/>
      <xsl:with-param name="p" select="$p + 1"/>
    </xsl:call-template>
```

```
  </xsl:when>

  <xsl:otherwise>
      <xsl:call-template name="reduce">
      <xsl:with-param name="x" select="2 *($x - floor($x))"/>
      <xsl:with-param name="n" select="16"/>
      <xsl:with-param name="p" select="$p"/>
    </xsl:call-template>
  </xsl:otherwise>

  </xsl:choose>

</xsl:template>

<!-- user-defined template:  reduce -->

<xsl:template name="reduce">
  <xsl:param name="x"/>
  <xsl:param name="n"/>
  <xsl:param name="p"/>

  <xsl:if test="$p = 0">.</xsl:if>


  <xsl:if test="$n &gt; 0">
    <xsl:value-of select="floor($x)"/>

    <xsl:call-template name="reduce">
      <xsl:with-param name="x" select="2 * ($x - floor($x))"/>
      <xsl:with-param name="n" select="$n - 1"/>
      <xsl:with-param name="p" select="$p - 1"/>
    </xsl:call-template>

  </xsl:if>

</xsl:template>

<!-- end of user-defined templates-->

</xsl:stylesheet>
```

The code employs two named templates. Template *normalize* repeatedly multiplies a given number by 0.5 until it has become less than 1. Each time a multiplication is carried out, the counter p increases by one unit. At the second stage, the template *reduce* produces the binary string of the reduced number, while the binary point is printed at an appropriate time.

Opening the *binary.xml* file with a *web* browser generates the following display:

```
The binary representation of 414.341 is: 110011110.0101011
The binary representation of 0.191 is: .0011000011100101
```

The number of binary digits printed, in this case 16, is specified as a parameter in the code. Alternatively, this number could have been provided as a datum in the *xml* file.

4.2.4 Hexadecimal representation of any number

An algorithm similar to that discussed in Section 4.2.3 for decimal-to-binary conversion can be developed for decimal-to-hexadecimal conversion. The hexadecimal representation employs the following sixteen hexadecimal digits:

```
0 1 2 3 4 5 6 7 8 9 A B C D E F
```

where the characters `A`–`F` represent the numbers 10–15. The decimal value of the hexadecimal number

$$\big(h_k h_{k-1} \cdots h_0 . h_{-1} h_{-2} \cdots h_{-l}\big)_{16}$$

is

$$h_k \times 16^k + h_{k-1} \times 16^{k-1} + \cdots + h_0 \times 16^0 + h_{-1} \times 16^{-1} + h_{-2} \times 16^{-2} + \cdots + h_{-l} \times 16^{-l},$$

where h_i are hexadecimal digits and k and l are two positive integers.

The conversion algorithm is based on two observations: each time a number is multiplied by 16, the hexadecimal point is shifted to the left by one place; each time a number is multiplied by $1/16 = 0.0625$ (divided by 16), the hexadecimal point is shifted to the right by one place.

These observations provide us with a practical method of deducing the hexadecimal representation of a number by a slight modification of the algorithm discussed in Section 4.2.3: (*a*) the given number is repeatedly divided by 16 until it becomes less than 1, (*b*) the hexadecimal representation of the transformed number is found, and (*c*) the hexadecimal point is shifted to the right by the number of divisions carried out.

As an example, we consider numbers contained in the following *xml* file entitled *hexa.xml*:

```
<?xml version="1.0" encoding="ISO-8859-1"?>
<?xml-stylesheet type="text/xsl" href="binary.xsl"?>
<anastasios>
```

```
   <number>414.341</number>
   <number>0.191</number>

</anastasios>
```

The name of the root element is **anastasios**. The accompanying *hexa.xsl* file implementing the conversion algorithm reads:

```
<xsl:stylesheet version="1.0"
      xmlns:xsl="http://www.w3.org/1999/XSL/Transform">

<!-- compute the hexadecimal representation
    of a number with n hexadecimal digits -->

<xsl:template match="anastasios">
<html>

<xsl:for-each select="number">
  The hexadecimal representation of <xsl:value-of select="."/> is:
  <xsl:call-template name="normalize">
    <xsl:with-param name="x" select="."/>
    <xsl:with-param name="p" select="0"/>
  </xsl:call-template>
  <br/>
</xsl:for-each>

</html>
</xsl:template>

<!-- user-defined template:  normalize -->

<xsl:template name="normalize">
  <xsl:param name="x"/>
  <xsl:param name="p"/>

  <xsl:choose>

  <xsl:when test="$x &gt; 1.0">
    <xsl:call-template name="normalize">
      <xsl:with-param name="x" select="0.0625 * $x"/>
      <xsl:with-param name="p" select="$p + 1"/>
    </xsl:call-template>
  </xsl:when>

  <xsl:otherwise>
    <xsl:call-template name="reduce">
      <xsl:with-param name="x" select="16 *($x - floor($x))"/>
      <xsl:with-param name="n" select="18"/>
```

```
      <xsl:with-param name="p" select="$p"/>
    </xsl:call-template>
  </xsl:otherwise>

</xsl:choose>

</xsl:template>

<!-- user-defined template:  reduce -->

<xsl:template name="reduce">
  <xsl:param name="x"/> <xsl:param name="n"/>
  <xsl:param name="p"/>

  <xsl:if test="$p = 0">.</xsl:if>

  <xsl:if test="$n &gt; 0">

    <xsl:variable name="spanakopita" select="floor($x)"/>

    <xsl:choose>

      <xsl:when test="$spanakopita &lt; 10">
        <xsl:value-of select="$spanakopita"/>
      </xsl:when>
      <xsl:otherwise>
        <xsl:variable name="tiropita" select="$spanakopita - 10"/>
        <xsl:value-of
          select="translate($tiropita, '012345', 'ABCDEF')"/>
      </xsl:otherwise>

    </xsl:choose>

    <xsl:call-template name="reduce">
      <xsl:with-param name="x" select="16 * ($x - floor($x))"/>
      <xsl:with-param name="n" select="$n - 1"/>
      <xsl:with-param name="p" select="$p - 1"/>
    </xsl:call-template>

  </xsl:if>

</xsl:template>

<!-- end of user-defined templates-->

</xsl:stylesheet>
```

The code employs two named functions (templates). Template *normalize* repeatedly multiplies a given number by $1/16 = 0.0625$ until it has become less than 1. Each time a multiplication is carried out, the counter `p` increases by one unit. At the second stage, the template *reduce* produces the hexadecimal string, while the hexadecimal point is printed at an appropriate time. The *xpath* function *translate* is used to convert numbers in the range 10–15 to corresponding digits, `A`–`F`.

Opening the *powi.xml* file with a *web* browser generates the display:

```
The hexadecimal representation of 414.341 is: 19E.574BC6A7EFA0000
The hexadecimal representation of 0.191 is: .30E5604189374C0000
```

Straightforward modifications of the code allow us to produce the representation of a number for any specified radix.

Exercise

4.2.1 *Base-17 representation*

Write an *xsl* code that produces the base-17 representation of an arbitrary real number.

4.3 Further applications of Xslt templates

To illustrate further the usefulness of *xsl* templates in scientific computing, we discuss the computation of integral powers and sums of powers of integers or real numbers.

4.3.1 *Integral power of a number*

We want to compute the power of a given positive or negative real number, a,

$$b = a^n,$$

where n is a positive integer exponent. Our strategy is to compute the power a^n by calculating a sequence of powers,

$$a, \quad a^2, \quad a^3, \quad \ldots \quad a^p, \quad \ldots \quad a^n. \tag{4.1}$$

The data contained in the following file entitled *powi.xml* define the base, a, and the exponent, n:

```
<?xml version="1.0" encoding="ISO-8859-1"?>
<?xml-stylesheet type="text/xsl" href="powi.xsl"?>
<onoufrios>

  <powint>
```

```
    <base>3.0</base>
    <exponent>3</exponent>
  </powint>

</onoufrios>
```

The name of the root element is `onoufrios`. The associated *xsl* file *powi.xsl* implementing the algorithm, referenced in the processing instruction stated in the second line, reads:

```
<xsl:stylesheet version="1.0"
     xmlns:xsl="http://www.w3.org/1999/XSL/Transform">

<!-- compute the integral power of a real number -->

<xsl:template match="onoufrios">
<html>

  <xsl:for-each select="powint">
    <xsl:value-of select="base"/>^<xsl:value-of select="exponent"/>=
    <xsl:call-template name="raise">
      <xsl:with-param name="base" select="base"/>
      <xsl:with-param name="exponent" select="exponent"/>
      <xsl:with-param name="power" select="base"/>
    </xsl:call-template>
  </xsl:for-each>

</html>
</xsl:template>

<!-- user-defined template:  raise -->

<xsl:template name="raise"> <xsl:param name="base"/>
          <xsl:param name="exponent"/> <xsl:param name="power"/>

  <xsl:choose>

    <xsl:when test="$exponent &gt; 1">
      <xsl:call-template name="raise">
        <xsl:with-param name="base" select="$base"/>
        <xsl:with-param name="exponent" select="$exponent - 1"/>
        <xsl:with-param name="power" select="$power * $base"/>
      </xsl:call-template>
    </xsl:when>

    <xsl:otherwise>
      <xsl:value-of select="$power"/>
    </xsl:otherwise>
```

```
  </xsl:choose>

</xsl:template>

<!-- end of user-defined template -->

</xsl:stylesheet>
```

Opening the *powi.xml* file with a *web* browser produces the expected display:

3.0^3= 27

The *xsl* implementation employs a user-defined template named *raise* to multiply the current power with the base, and thus produce the next intermediate power. The values of the parameters inside template are accessed by prepending the dollar sign ($) to their names. Each time a member of the sequence (4.1) is computed, the parameter `exponent` is reduced by one unit. The computations terminate when `exponent` has become zero.

If we print the exponent inside the root *xsl* template referring to the root *xml* element `onoufrios` after the `raise` template has been applied, we will obtain the original value specified in the *xml* document, in this case 3. This observation clearly demonstrates that *exponent* becomes a local variable when passed to the template.

Negative integer exponents

When the exponent n is a negative integer, we may use the identity $a^n = 1/a^{-n}$ to compute and then invert by division a positive integral power. Alternatively, but less efficiently, we may replace multiplication by division in the code.

Comparison with C++

A C++ function named *powr.cc* playing the role of a named template that computes the power a^n is listed below for comparison:

```
double powr (double a, int n)
{
   int k;
   double accum = 1.0;
   for(k=1; k<=n; k++)
   {
   accum = accum * a;
   }
return accum;
}
```

Real numbers registered in double precision (`double`) and integers (`int`) are employed as variables. The recursive use of functions (templates) is not necessary, thanks to our ability to update an accumulator. The compilation and execution of a C++ code was discussed in Section 1.6.3.

Comparing the *xsl* code with the equivalent C++ code confirms our well founded suspicion, now turned into conviction, that C, C++, *fortran*, or any other high-level scientific language is more efficient than *xsl*. In our example, the *xsl* implementation is restricted by the absence of the counterpart of the `for` loop and the inability to perform direct variable update.

4.3.2 Highest integer with a given number of bits

The highest integer that can be described with a specified number of bits, n, is given by

$$2^0 + 2^1 + \cdots + 2^{n-1} = 2^n - 1.$$

For example, when $n = 2$, the maximum integer is $3 = (11)_2$. The following file *bits.xml* defines the number of available bits:

```
<?xml version="1.0" encoding="ISO-8859-1"?>
<?xml-stylesheet type="text/xsl" href="bits.xsl"?>
<touloumba>

  <nbits>16</nbits>

</touloumba>
```

The name of the root element is `touloumba`. The associated *xsl* file entitled *bits.xsl* reads:

```
<xsl:stylesheet version="1.0"
      xmlns:xsl="http://www.w3.org/1999/XSL/Transform">
<xsl:template match="touloumba">
<html>

<!-- compute the maximum integer described by specified bits -->

<xsl:for-each select="nbits">

  <table cellpadding="1"> <tr align="center">
      <th>bits-</th><th>-increment-</th>
      <th>-largest integer</th></tr>

  <xsl:call-template name="raise_add">
    <xsl:with-param name="nbits" select="."/>
    <xsl:with-param name="counter" select="1"/>
    <xsl:with-param name="increment" select="1"/>
    <xsl:with-param name="sum" select="1"/>
```

```
  </xsl:call-template>

  </table>

  </xsl:for-each>

</html>
</xsl:template>

<!-- user-defined template:  raise_add -->

<xsl:template name="raise_add"> <xsl:param name="nbits"/>
    <xsl:param name="counter"/> <xsl:param name="increment"/>
    <xsl:param name="sum"/>

  <xsl:if test="not($counter &gt; $nbits)">
    <tr align="center">
    <td><xsl:value-of select="$counter"/></td>
    <td><xsl:value-of select="$increment"/></td>
    <td><xsl:value-of select="$sum"/></td></tr>
  </xsl:if>

  <xsl:if test="$counter &lt; $nbits">
    <xsl:call-template name="raise_add">
      <xsl:with-param name="nbits" select="$nbits"/>
      <xsl:with-param name="counter" select="$counter + 1"/>
      <xsl:with-param name="increment" select="2 * $increment"/>
      <xsl:with-param name="sum" select="$sum + 2 * $increment"/>
    </xsl:call-template>

  </xsl:if>

</xsl:template>

<!-- end of user-defined template:  raise_add -->

</xsl:stylesheet>
```

Opening the *bits.xml* file with a the *web* browser produces the display:

```
bits--increment--largest integer

        1     1     1
        2     2     3
        3     4     7
        4     8    15
        5    16    31
        6    32    63
        7    64   127
```

```
 8   128   255
 9   256   511
10   512  1023
11  1024  2047
12  2048  4095
13  4096  8191
14  8192 16383
15 16384 32767
16 32768 65535
```

The method implemented in the *bits.xsl* code is based on an algorithm similar to that used for computing the power of a number, as discussed in Section 4.3.1.

Fortran code

It is of interest to compare the *xsl* code with an equivalent *fortran* code contained in the following file named *bits.f*:

```
      program bits

c----------------------
c compute the maximum integer that can be described
c with a specified number of bits
c----------------------

      Implicit Double Precision (a-h,o-z)
      Integer p,q

      write (6,*) " Will compute the greatest integer "
      write (6,*) " that can be described with n bits "
 98   write (6,*) " Enter the number of bits (less than 32)"
      write (6,*) " 0 to quit "
      write (6,*) " ----------"
      read (5,*) n

      If(n.eq.0) Go to 99

      write (6,101)

      q = 0

      Do i=0,n-1
        p = 2**i
        q = q+p
        write (6,100) i+1,p,q
      End Do

      Go to 98   !  return to repeat
```

```
99    Continue      !  done

100   Format (1x,i5,2(1x,i15))
101   Format (" bits",5x,"increment",5x,"highest integer")

      Stop
      End
```

The clarity and elegance of the *fortran* code are noteworthy. To compile the code and generate an executable named *bits_f*, we launch the *fortran* compiler (f77) by issuing the following statement in a terminal (command-line window):

```
f77 -o bits_f bits.f
```

To run the executable, we issue the statement:

```
./bits_f
```

and then hit the ENTER key. A typical session follows:

```
 Will compute the greatest integer
 that can be described with n bits
 Enter the number of bits (less than 32)
 0 to quit
 ---------
 6

 bits          increment          highest integer
   1               1                    1
   2               2                    3
   3               4                    7
   4               8                   15
   5              16                   31
   6              32                   63
Enter the number of bits (less than 32)
0 to quit
---------
0
```

C++ code

An equivalent self-contained C++ code residing in the file *bits.cc* reads:

```
/* ---------------- ---------------------------
Greatest integer that can be described by n bits
-----------------------------------------------*/

#include <iostream>
```

```
#include <iomanip>
#include <cmath>

using namespace std;

/* -------- main program -------- */

int main()
{
int n=1;
int i;
const int two = 2;
cout << " Will compute the greatest integer\n";
cout << " that can be described with n bits\n";

  while(n!=0)
  {
  cout << "\n";
  cout << " Please enter the number of bits\n";
  cout << " (should be less than 32)\n";
  cout << " q to quit\n";
  cout << " -----------\n";

  if(!(cin >> n)) break;

  cout << setw(13) << " bits " << " "
      << setw(10) << " increment" << " "
        << setw(16) << "highest integer" << "\n";

  int q = 0;

  for (i=0; i<=n-1; i++)
    {
    int p = pow(two, i);
    q = q + p;
    cout << setw(10) << i+1 << " " << setw(10) << p << " "
      << setw(10) << q << "\n";
    };

  };

return 0;
}
```

Since the mathematical function *pow* is used, the headers of the mathematical library (*cmath*) have been included at the beginning of the code. To compile the code and generate an executable named *bits_cc*, we run the C++ compiler (g++) by issuing the following statement in a command-line window:

```
g++ -o bits_cc bits.cc
```

To run the executable, we type:

```
./bits_cc
```

and then hit the ENTER key. A typical session follows:

```
Will compute the greatest integer
that can be described with n bits

Please enter the number of bits
       (should be less than 32)
q to quit
-----------
5
      bits       increment   highest integer
        1            1             1
        2            2             3
        3            4             7
        4            8            15
        5           16            31

Please enter the number of bits
       (should be less than 32)
q to quit
-----------
q
```

With regard to clarity, among *xsl*, *fortran*, and C++, *fortran* is the clear winner. This explains why *fortran* is still a popular programming language among scientists and engineers in core applications where advanced structures and memory management are not required.

Exercises

4.3.1 *Negative power of a real number*

Write an *xsl* code that computes and prints the negative integral power of an arbitrary real number.

4.3.2 *Power of a complex number*

Write an *xsl* code that computes and prints the integral power of an arbitrary complex number defined by its real and imaginary parts.

4.3.3 *Power of a matrix*

(*a*) Write an *xml* file that contains a 2×2 real matrix, and an *xsl* file that computes and prints the power of this matrix. (*b*) Repeat for a complex matrix.

4.3.4 *Factorial*

Write an *xml* file that defines an integer, and a companion *xsl* file that computes the factorial of the integer. The factorial of an integer, m, is $m! = 1 \cdot 2 \cdots m$, subject to the convention that $0! = 1$.

4.4 Square root of a number

A program can be written to compute the square root of a specified real positive number, a,

$$b = \sqrt{a}.$$

The number a is the radical and the number b is the square root of the radical. Using Newton's method, we guess the value of b, and then improve our guess by computing a new estimate,

$$b^{\text{new}} = \frac{1}{2}\,(b + \frac{a}{b}).$$

The iterations terminate when the new number is equal to the old number within a specified tolerance.

The following file named *sqrt.xml* defines several radicals, a:

```
<?xml version="1.0" encoding="ISO-8859-1"?>
<?xml-stylesheet type="text/xsl" href="sqrt.xsl"?>
<rodakino>

  <number>3.0</number>
  <number>3.4</number>
  <number>4.0</number>

</rodakino>
```

The name of the root element is `rodakino`, which means *peach* in Greek. The associated *xsl* file entitled *sqrt.xsl* implementing Newton's method with 8 iterations reads:

```
<xsl:stylesheet version="1.0"
      xmlns:xsl="http://www.w3.org/1999/XSL/Transform">

<!-- compute the square root of a number
    with a specified number of iterations -->

<xsl:template match="sqrt">
<html>

  <xsl:for-each select="number">
    <xsl:call-template name="update">
```

```
      <xsl:with-param name="radical" select="."/>
      <xsl:with-param name="root" select=". div 2"/>
      <xsl:with-param name="iteration" select="8"/>
    </xsl:call-template>
  </xsl:for-each>

</html>
</xsl:template>

<!-- user-defined template:  update -->

<xsl:template name="update"> <xsl:param name="radical"/>
      <xsl:param name="root"/> <xsl:param name="iteration"/>

  <xsl:choose>

    <xsl:when test="$iteration &gt; 0">
      <xsl:call-template name="update">
        <xsl:with-param name="radical" select="$radical"/>
        <xsl:with-param name="root"
          select="($root + $radical div $root) div 2"/>
        <xsl:with-param name="iteration" select="$iteration - 1"/>
      </xsl:call-template>
    </xsl:when>

    <xsl:otherwise>
      <xsl:value-of select="$root"/><br/>
    </xsl:otherwise>

  </xsl:choose>
</xsl:template>

<!-- end of user-defined function:  update -->

</xsl:stylesheet>
```

Opening the *sqrt.xml* file with a the *web* browser produces the display:

```
1.7320508075688772
1.8439088914585775
2
```

representing accurate or exact approximations to the square roots of 3.0, 3.4, and 4.0, respectively.

A judicious initial guess is implemented in the code, $b = \frac{1}{2}\, a$. Each time the template *update* is applied, the number of remaining iterations is reduced by one unit. The calculations terminate when the number of remaining iterations has become zero. At that stage, the computed square root is printed on the screen. Note that the variable *iteration* is local to the *update* template and will have the initial value of 8 if printed inside the root template of the *xsl* code following the computation of the square root.

Number of iterations as a global variable

The number of iterations could have been accommodated into a variable named `niter` by the statement:

```
<xsl:variable name="niter" select="8"/>
```

The variable `niter` is global if this statement is placed immediately after the `stylesheet` declaration in the *xsl* code, and local if the statement is placed inside the root template. We may then use the line:

```
<xsl:with-param name="iteration" select="$niter"/>
```

when calling the template `update` from the root template.

Terminating at a specified accuracy

The *xsl* code can be modified so that the iterations terminate when a specified level of accuracy has been reached, subject to a maximum number of iterations to prevent runoff. The new data are read from the following file entitled *sqrt1.xml*:

```
<?xml version="1.0" encoding="ISO-8859-1"?>
<?xml-stylesheet type="text/xsl" href="sqrt1.xsl"?>
<giouvarlakia>

  <number>2.1</number>
  <number>4.1</number>
  <number>16.1</number>

</giouvarlakia>
```

The name of the root element is `giouvarlakia`. The improved *xsl* code contained in a file entitled *sqrt1.xsl* reads:

```
<xsl:stylesheet version="1.0"
      xmlns:xsl="http://www.w3.org/1999/XSL/Transform">

<!-- compute the square root of a number with specified accuracy -->

<xsl:variable name="accuracy" select="0.00000001"/>

<xsl:template match="giouvarlakia">
```

```
<html>

  <xsl:for-each select="number">
    <xsl:call-template name="update">
      <xsl:with-param name="radical" select="."/>
      <xsl:with-param name="root" select=". div 2"/>
      <xsl:with-param name="iteration" select="5"/>
      <xsl:with-param name="error" select="1000.00"/>
    </xsl:call-template>
  </xsl:for-each>

</html>
</xsl:template>

<!-- user-defined template:  update -->

<xsl:template name="update"> <xsl:param name="radical"/>
      <xsl:param name="root"/> <xsl:param name="iteration"/>
      <xsl:param name="error"/>

<xsl:choose>

<xsl:when test="$error &lt; -$accuracy or $error &gt; $accuracy">

  <xsl:choose>

    <xsl:when test="$iteration &gt; 0">
      <xsl:call-template name="update">
        <xsl:with-param name="radical" select="$radical"/>
        <xsl:with-param name="root"
          select="($root + $radical div $root) div 2"/>
        <xsl:with-param name="iteration" select="$iteration - 1"/>
        <xsl:with-param name="error"
          select="(-$root + $radical div $root) div 2"/>
      </xsl:call-template>
    </xsl:when>

    <xsl:otherwise>
      The square root of <xsl:value-of select="$radical"/>
      could not be computed in 5 iterations <br/>
      <br/>
    </xsl:otherwise>

  </xsl:choose>

</xsl:when>

<xsl:otherwise>
    The square root of <xsl:value-of select="$radical"/>
```

```
    is <xsl:value-of select="$root"/>
    accurate to the eighth decimal place
    <br/>
</xsl:otherwise>

</xsl:choose>

</xsl:template>

<!-- end of user-defined template -->

</xsl:stylesheet>
```

The numerical error is defined as the difference between the current and previous estimates of the square root. In the algorithm, if the value of the parameter `error` is higher than a specified threshold in absolute value, another iteration is performed, provided that the number of iterations does not exceed a specified safety limit, which is set to 5 in the code. The double check necessitates the use of two nested `choose` structures, as shown in the code. The first check is implemented by the line

```
<xsl:when test="$error &lt; -$accuracy or $error &gt; $accuracy">
```

where `accuracy` is a global variable defined and evaluated in the sixth line of the *xsl* code. Alternatively, this variable could be evaluated as an appropriate *xml* input.

Exercises

4.4.1 *nth root of a number*

Write an *xsl* code that produces the nth root of a real and positive number, a, denoted by b, defined so that $b^n = a$ where n is a positive integer. The algorithm should be based on Newton's formula,

$$b^{\text{new}} = \frac{1}{n}\left((n-1)\,b + \frac{a}{b^{n-1}} \right). \tag{4.1}$$

A suitable initial guess should be made.

4.4.2 *Division by multiplication*

Write an *xsl* code that produces the inverse of a number, a, denoted by $b = 1/a$, based on the recursive formula

$$b^{\text{new}} = b\,(2 - a\,b). \tag{4.2}$$

A suitable initial guess should be made.

4.4.3 *Newton's method*

Newton's method is used to find a zero of an algebraic equation, $f(x) = 0$; for example $f(x) = e^x - 4.5$. An initial guess is made and then improved by iteration. The following block of a *fortran* code implements the method:

```
      Do i=1,Niter
        call fun(x,f) !  function evaluation
        x1 = x + eps !  derivative by finite differences
        call fun(x1,f1)
        Df = (f1-f)/eps
        Dx = - f/Df
        x = x+Dx
        if(abs(Dx).le.tol) Go to 99
      End Do

 99   Continue !  Done
```

where the companion function `fun` evaluates $f(x)$, and `eps` is a small number used to approximate the derivative $f'(x)$ by numerical differentiation. Write an *xml* code that implements Newton's method for a function $f(x)$ of your choice.

4.5 Exponential of a number

The exponential of a given number, x, can be computed from its Maclaurin series expansion,

$$e^x = 1 + x + \frac{1}{2!}\,x^2 + \frac{1}{3!}\,x^3 + \cdots + \frac{1}{m!}\,x^m + \cdots, \tag{4.1}$$

where $m! = 1 \cdot 2 \cdots m$ is the factorial. The series converges for any value of x.

The Maclaurin series can be summed conveniently by iteration using Horner's algorithm. If only four terms are kept, the sum can be rearranged into a nested product,

$$e^x \simeq 1 + x\left(1 + \frac{1}{2}\,x\left(1 + \frac{1}{3}\,x\right)\right). \tag{4.2}$$

In practice, we compute the unfolded recursive sequence

$$w_1 = 1 + \frac{1}{3}\,x, \qquad w_2 = 1 + \frac{1}{2}\,x\,w_1, \qquad w_3 \simeq 1 + x\,w_2 \simeq e^x. \tag{4.3}$$

More generally, we compute a sequence of n numbers, $w_1, w_2, \ldots, w_{n-1}, w_n$, leading us to an approximation of the exponential,

$$w_1 = 1 + \frac{1}{n}\,x, \quad w_2 = 1 + \frac{1}{n-1}\,x\,w_1, \quad \ldots,$$
$$w_{n-1} = 1 + \frac{1}{2}\,x\,w_{n-2}, \qquad w_n = 1 + x\,w_{n-1} \simeq e^x, \tag{4.4}$$

where n is a specified truncation level determining the accuracy of the final result.

In our application, an assortment of numbers whose exponentials are sought is defined in the following file entitled *exp.xml*:

```
<?xml version="1.0" encoding="ISO-8859-1"?>
<?xml-stylesheet type="text/xsl" href="exp.xsl"?>
<diogenes>

  <number>0.34</number>
  <number>0.8</number>
  <number>-0.5</number>

</diogenes>
```

The name of the root element is `diogenes`. The accompanying *exp.xsl* file reads:

```
<xsl:stylesheet version="1.0"
      xmlns:xsl="http://www.w3.org/1999/XSL/Transform">

<!-- compute the exponential of a number from the Maclaurin series
  expansion keeping "nterms" terms -->

<xsl:variable name="nterms" select="8"/>

<xsl:template match="diogenes">
<html>

<xsl:for-each select="number">
  e^<xsl:value-of select="."/> =
  <xsl:call-template name="nested">
    <xsl:with-param name="x" select="."/>
    <xsl:with-param name="e" select="1 + ( .  div $nterms)"/>
    <xsl:with-param name="n" select="$nterms"/>
  </xsl:call-template>
</xsl:for-each>

</html>
</xsl:template>

<!-- user-defined template:  nested -->

<xsl:template name="nested"> <xsl:param name="x"/>
        <xsl:param name="e"/> <xsl:param name="n"/>

  <xsl:choose>
```

```
    <xsl:when test="$n &gt; 1">
      <xsl:call-template name="nested">
        <xsl:with-param name="x" select="$x"/>
        <xsl:with-param name="e" select="1 + $x * $e div ($n - 1)"/>
        <xsl:with-param name="n" select="$n - 1"/>
      </xsl:call-template>
    </xsl:when>

    <xsl:otherwise>
      <xsl:value-of select='format-number($e,"##.######")'/>
      (computed from the Maclaurin series with
      <xsl:value-of select="$nterms"/> terms)<br/>
    </xsl:otherwise>

    </xsl:choose>

  </xsl:template>

  <!-- end of user-defined template -->

  </xsl:stylesheet>
```

Opening the file *exp.xml* with a *web* browser produces the display:

```
e^0.34 = 1.404948 (computed from the Maclaurin series with 8 terms)
e^0.8 = 2.225541 (computed from the Maclaurin series with 8 terms)
e^-0.5 = .606531 (computed from the Maclaurin series with 8 terms)
```

The number of terms retained, `nterms`, is declared and evaluated as a global variable in the seventh line of the *xsl* code. The template *nested* is then recursively called to implement Horner's algorithm. The value of `nterms` is accessed by the template `nested` and printed at the conclusion.

Large arguments

An increasing number of terms must be retained as the magnitude of x becomes higher. To circumvent this difficulty, we write

$$e^x = \left(e^{x/p}\right)^p,$$

where p is a positive integer chosen so that the magnitude of the inner exponent is less than unity, $|x/p| < 1$.

Assume that the exponent, x, is positive. If $0 \le x \le 1$, we choose $p = 1$; if $1 < x \le 2$, we choose $p = 2$; if $2 < x \le 3$, we choose $p = 3$. More generally, p is the floor of x. The method proceeds by first computing the exponential $c = e^y$, where $y = x/p$, and then computing the integral power $e^x = c^p$ using the algorithm discussed in this section. The integral power of a number can be computed using the method discussed in Section 4.3.1.

Negative arguments

If x is negative, we use the identity $e^x = 1/e^{|x|}$ to compute and then invert the exponential of a positive number in the denominator.

Exercises

4.5.1 *Exponential of an arbitrary number*

Implement the methods discussed in the text into an *xsl* code that produces the exponential of an arbitrary positive or negative real number, x.

4.5.2 *Trigonometric functions*

(a) The Maclaurin series of the sine function is

$$\sin x = x - \frac{1}{3!}\,x^3 - \frac{1}{5!}\,x^5 + \frac{1}{7!}\,x^7 + \cdots . \tag{4.5}$$

Develop a method of computing the sine of a real number, x, in the interval $[0, \pi/2)$ based on Horner's algorithm. Generalize the method to arbitrary values of x based on the properties

$$\sin x = \sin(n\pi - x), \qquad \sin x = \sin(x + m\pi), \tag{4.6}$$

where n is an odd integer and m is an even integer.

(b) Use the algorithm described in (a) to compute the cosine of an arbitrary real number based on the identity $\cos x = \sin(x + \pi/2)$.

4.6 Natural logarithm of a number

The natural logarithm of a given positive number, x, can be computed from the Taylor series expansion of the logarithmic function about unity, yielding

$$\ln x = t - \frac{1}{2}\,t^2 + \frac{1}{3}\,t^3 - \frac{1}{4}\,t^4 + \cdots , \tag{4.1}$$

where $t = x - 1$, provided that $0 < x < 2$ so that $-1 < t < 1$. For values $x \geq 2$, the Taylor series does not converge. If only four terms are kept in the Taylor series, the sum can be rearranged into a nested product,

$$\ln x = t\left(1 - \frac{1}{2}\,t\,(1 - \frac{2}{3}\,t\,[1 - \frac{3}{4}\,t\,])\right), \tag{4.2}$$

which suggest computing the recursive sequence

$$w_1 = 1 - \frac{3}{4}\,t, \qquad w_2 = 1 - \frac{2}{3}\,t\,w_1, \qquad w_3 = 1 - \frac{1}{2}\,t\,w_2, \qquad \ln x \simeq t\,w_3. \tag{4.3}$$

More generally, we compute the sequence

$$w_1 = 1 - \frac{n-1}{n}\,t, \quad w_2 = 1 - \frac{n-2}{n-1}\,t w_1, \quad \ldots,$$
$$w_{n-1} = 1 - \frac{1}{2}\,t w_{n-2}, \qquad \ln x \simeq t w_{n-1}, \qquad (4.4)$$

where n is a specified integer. The accuracy of the computation improves as n becomes increasingly large.

In our application, an assortment of numbers whose natural logarithms are sought are defined in the following *xml* file entitled *ln.xml*:

```
<?xml version="1.0" encoding="ISO-8859-1"?>
<?xml-stylesheet type="text/xsl" href="ln.xsl"?>
<patsavoura>

   <number>0.14</number>
   <number>1.73</number>

</patsavoura>
```

The name of the root element is `patsavoura`. The accompanying *ln.xsl* file reads:

```
<xsl:stylesheet version="1.0"
      xmlns:xsl="http://www.w3.org/1999/XSL/Transform">

<!-- natural logarithm of a number by Taylor series expansion about 1
keeping "nterms" (even) terms -->

<xsl:variable name="nterms" select="32"/>

<xsl:template match="themistocles">
<html>

<xsl:for-each select="number">
  ln(<xsl:value-of select="."/>) =
    <xsl:call-template name="nested">
      <xsl:with-param name="t" select=".  - 1"/>
      <xsl:with-param name="ln"
        select="1 - ($nterms - 1) div $nterms*(.  -1 )"/>
      <xsl:with-param name="n" select="$nterms"/>
    </xsl:call-template>
</xsl:for-each>

</html>
</xsl:template>
```

```
<!-- user-defined template:  nested -->

<xsl:template name="nested">
  <xsl:param name="t"/>
  <xsl:param name="ln"/>
        <xsl:param name="n"/>

<xsl:choose>

  <xsl:when test="$n &gt; 2">
  <xsl:call-template name="nested">
    <xsl:with-param name="t" select="$t"/>
    <xsl:with-param name="ln"
      select="1 - $t*$ln*($n - 2) div ($n - 1)"/>
    <xsl:with-param name="n" select="$n - 1"/>
  </xsl:call-template>
  </xsl:when>

  <xsl:otherwise>
    <xsl:value-of select='format-number($t*$ln,"##.######")'/>
    (computed from the Taylor series with
    <xsl:value-of select="$nterms"/> terms)<br/>
  </xsl:otherwise>

</xsl:choose>
</xsl:template>

<!-- end of user-defined template -->

</xsl:stylesheet>
```

Opening the file *ln.xml* with a *web* browser generates the display:

```
ln(0.14) = -1.964826 (computed from the Taylor series with 32 terms)
ln(1.73) = .548121 (computed from the Taylor series with 32 term
```

The logarithm of a number that is less than 1 is negative, whereas the logarithm of a number that is higher than 1 is positive.

Large and small arguments

The Taylor series converges slowly when $1.5 < x < 2$ and diverges when $x \geq 2$. To ensure rapid convergence, we write

$$x = a^p \xi, \tag{4.5}$$

where $a > 1$ is an arbitrary constant and the integer p is adjusted so that $0 < \xi < 1.5$. Using the properties of the logarithm, we find that

$$\ln x = p \ln a + \ln \xi. \tag{4.6}$$

The logarithm $\ln \xi$ is computed using the method discussed in this section. It is convenient to set $a = \mathrm{e} \equiv 2.7182818284\cdots$ so that $\ln a = 1$. A similar method can be implemented when x is too close to zero.

Exercises

4.6.1 *Logarithm of an arbitrary number*

Implement the method discussed in the text into an *xsl* code that computes the logarithm of an arbitrary positive number.

4.6.2 *Logarithm with arbitrary base*

Write a code that produces the base-q logarithm of an arbitrary positive number, where q is an arbitrary base.

4.7 Recursive sequences

We are interested in computing a recursive sequence of numbers, where the next number, x_{n+1}, is a linear combination of the current and one previous number, x_n and x_{n-1},

$$x_{n+1} = a\,x_n + b\,x_{n-1}, \tag{4.1}$$

where a and b are two specified coefficients. The first two members of the sequence, x_1 and x_2, are provided.

In our application, the two starting numbers are contained in the following *xml* file entitled *recursive.xml*:

```
<?xml version="1.0" encoding="ISO-8859-1"?>
<?xml-stylesheet type="text/xsl" href="recursive.xsl"?>
<ouzo>

  <number1>0</number1>
  <number2>1</number2>

</ouzo>
```

The name of the root element is `ouzo`. The accompanying *recursive.xsl* file reads:

```
<xsl:stylesheet version="2.0"
     xmlns:xsl="http://www.w3.org/1999/XSL/Transform">
<xsl:variable name="a" select="1.0"/>
<xsl:variable name="b" select="1.0"/>
<xsl:template match="ouzo">
<html>
```

```
  <xsl:value-of select="number1"/>

  <xsl:call-template name="move">
    <xsl:with-param name="n1" select="number1"/>
    <xsl:with-param name="n2" select="number2"/>
    <xsl:with-param name="nterms" select="16"/>
  </xsl:call-template>

</html>
</xsl:template>

<!-- user-defined template:  move -->

<xsl:template name="move">
  <xsl:param name="n1"/>
  <xsl:param name="n2"/>
  <xsl:param name="nterms"/>

<xsl:if test="$nterms > 1">
    &#183; <xsl:value-of select="$n2"/>

    <xsl:call-template name="move">
      <xsl:with-param name="n1" select="$n2"/>
      <xsl:with-param name="n2" select="$a * $n1 + $b * $n2"/>
      <xsl:with-param name="nterms" select="$nterms - 1"/>
    </xsl:call-template>

</xsl:if>

</xsl:template>

<!-- end of user-defined template -->

</xsl:stylesheet>
```

Opening the *recursive.xml* file with a *web* browser prints the first sixteen members of the Fibonacci series:

```
0 - 1 - 1 - 2 - 3 - 5 - 8 - 13 - 21 - 34 - 55 - 89 - 144 - 233 - 377 - 610
```

The algorithm employs the countdown parameter `nterms` and two other parameters, `n1` and `n2`, holding the last available pair.

Exercise

4.7.1 *Recursive product*
Consider the recursive product $x_{n+1} = a\, x_n x_{n-1}$, where a is a given coefficient

and the two initial values, x_1 and x_2, are provided. Modify the *xsl* code listed in the text into a program that computes this recursive sequence and then use the program to compute the integral power of a real number.

4.8 Greatest common divisor of two integers

The greatest common divisor (gcd) of two integers is defined as the maximum integer that divides both. The greatest common divisor can be found using Euclid's algorithm according to the following steps: (*a*) compute the difference between the large and small integer, (*b*) replace the large integer by the difference, (*c*) if the two numbers are the same, stop; otherwise repeat the procedure.

The following *xml* file entitled *gcd.xml* defines two integers:

```
<?xml version="1.0" encoding="ISO-8859-1"?>
<?xml-stylesheet type="text/xsl" href="gcd.xsl"?>
<gaidouraki>

  <pair>
    <int1>5183</int1>
    <int2>2414</int2>
  </pair>

</gaidouraki>
```

The name of the root element is `gaidouraki`. The associated *gcd.xsl* file implementing Euclid's algorithm reads:

```
<xsl:stylesheet version="2.0"
     xmlns:xsl="http://www.w3.org/1999/XSL/Transform">
<xsl:template match="gaidouraki">
<html>

  <xsl:for-each select="pair">

    The greatest common denominator of:  <xsl:value-of select="int1"/>
    and <xsl:value-of select="int2"/> is:

    <xsl:call-template name="update">
      <xsl:with-param name="in1" select="int1"/>
      <xsl:with-param name="in2" select="int2"/>
    </xsl:call-template>

  </xsl:for-each>

</html>
</xsl:template>
```

```
<!-- user-defined template: update -->

<xsl:template name="update">
  <xsl:param name="in1"/>
  <xsl:param name="in2"/>

  <xsl:choose>
  <xsl:when test="$in1 != $in2"> <!--the numbers are not equal-->

    <xsl:variable name="diff">
      <xsl:if test="$in1 &gt; $in2">
        <xsl:value-of select="$in1 - $in2"/>
      </xsl:if>
      <xsl:if test="$in1 &lt; $in2">
        <xsl:value-of select="$in2 - $in1"/>
      </xsl:if>
    </xsl:variable>

    <xsl:variable name="small">
      <xsl:if test="$in1 &gt; $in2">
        <xsl:value-of select="$in2"/>
      </xsl:if>
      <xsl:if test="$in1 &lt; $in2">
        <xsl:value-of select="$in1"/>
      </xsl:if>
    </xsl:variable>

    <xsl:call-template name="update">
      <xsl:with-param name="in1" select="$diff"/>
      <xsl:with-param name="in2" select="$small"/>
    </xsl:call-template>

  </xsl:when>

  <xsl:otherwise>       <!-- the numbers are equal -->
    <xsl:value-of select="$in1"/> <br/>
  </xsl:otherwise>

  </xsl:choose>

</xsl:template>

<!-- end of user-defined template -->

</xsl:stylesheet>
```

Opening the *gcd.xml* file with a *web* browser produces the display:

The greatest common divisor of: 5183 and 2414 is: 71

The *xsl* implementation employs the named template *update* that receives two integers and performs suitable calculations. The variables `diff` and `small` are first defined and evaluated, and the template calls itself recursively, each time replacing the first integer with `diff` and the second integer with `small`. The iterations terminate when these two numbers have become equal. The algorithm is yet another implementation of recursive function calling.

C++ code

For comparison, an equivalent C++ code complete with interactive input from a command-line window and output to a terminal is listed below:

```
#include <iostream>
using namespace std;

int main()
{
int n,m,k,nsave;

  cout<<"\n Will compute the Greatest Common Divisor";
  cout<<"\n\t of two positive integers\n";
  cout<<"\n Please enter the two integers";
  cout<<"\n\t0 for either one to quit\n";
  cin>>n; cin>>m;

while ( (n!=0) && (m!=0) )
    {
    if(n==m)
    { k=n;
      cout<<"\nThe Greatest Common Divisor is: "<<k<<"\n";
    }
else while (n!=m)
    {
      if(n>m) {
        nsave = m;
        m = n;
        n = nsave;
        }
      k=m-n; m=n; n=k;
    if(n==m) {
        k=n;    cout<<"\nThe Greatest Common Divisor is: "<<k<<"\n";
        }
      }

cout<<"\nPlease enter the two integers";
cout<<"\n\t0 for either one to quit\n";
cin>>n; cin>>m;
}
```

```
cout<<"\nThank you for your business.\n";
return 0;
}
```

The necessary input/output headers (*iostream*) have been included at the beginning of the code. The code consists of a main program where the use of the standard namespace is first specified. Four integers, `b,m,k`, and `nsave`, are then declared, and a dialog is initiated using the *cout* and *cin* output and input functions. We observe that functions (templates) are not needed to implement Euclid's algorithm in the C++ code.

To compile the C++ program and create an executable binary file named *gcd* in a terminal (command-line window), we run the C++ compiler by typing:

```
g++ -o gcd gcd.cc
```

and then hit the ENTER key. To run the executable, we type:

```
./gcd
```

and then hit the ENTER key. A typical session follows:

```
Will compute the Greatest Common Divisor
     of two positive integers

Please enter the two integers 0 for either one to quit
     5183 2414

The Greatest Common Divisor is: 71

Please enter the two integers
     0 for either one to quit
0 0

Thank you for your business.
```

The superiority of C++ is apparent.

Exercise

4.8.1 *Greatest common divisor in a language of your choice*

Implement Euclid's algorithm in a code written in a programming language of your choice.

4.9 Student roster

An *xml* book could not be complete without a student roster that can be used instead of a spreadsheet. Consider the following *xml* file entitled *roster.xml* containing student grades for two homework assignments (hw1 and hw2) and two exams (ex1 and ex2) in an engineering course:

```
<?xml version="1.0" encoding="ISO-8859-1"?>
<?xml-stylesheet type="text/xsl" href="grades.xsl"?>

<course name="ENG101" term="Spring 2005">

<hw1 max="100" weight="1.0"/>
<hw2 max="100" weight="1.0"/>
<ex1 max="60" weight="3.0"/>
<ex2 max="70" weight="4.0"/>

<student>
  <firstname>Jim</firstname>
  <middlename>R.</middlename>
  <lastname>Smith</lastname>
  <hw1 grade="80"/>
  <hw2 grade="70"/>
  <ex1 grade="30"/>
  <ex2 grade="30"/>
</student>

<student>
  <firstname>Kathryne</firstname>
  <middlename>L.</middlename>
  <lastname>Sotiris</lastname>
  <hw1 grade="90"/>
  <hw2 grade="59"/>
  <ex1 grade="55"/>
  <ex2 grade="55"/>
</student>

<student>
  <firstname>Anne</firstname>
  <middlename>R.</middlename>
  <lastname>Hildebrand</lastname>
  <hw1 grade="92"/>
  <hw2 grade="83"/>
  <ex1 grade="57"/>
  <ex2 grade="70"/>
</student>

</course>
```

The name of the root element is `course`. Attributes of the root element are the course name and academic term during which the course was taught. The maximum possible grade and weight of each homework or exam are specified as element attributes at the beginning of the file. A person with basic data-entry skills can easily modify this file to add more students, homework assignments, or exams.

On a scale from 0 to 100, the course grade will be computed using the formula

$$\text{course grade} = 100 \times \frac{\sum \text{grade} \times \text{weight}}{\sum(\text{maximum grade}) \times \text{weight}}.$$

A letter grade will then be assigned.

The accompanying *xsl* code contained in a file entitled *roster.xsl* counts the students using the intrinsic *xslt* functions *count* and *sum*, and produces a spreadsheet where the student data are printed in alphabetical order using the intrinsic *xslt* function *sort*:

```
<xsl:stylesheet version="1.0"
      xmlns:xsl="http://www.w3.org/1999/XSL/Transform">
<xsl:template match="course">
<html>

<!--
Student Grades Spreadsheet

  ns:  number of students in class
  hwXm:  homework maximum
  exXm:  exam maximum
  hwXw:  homework weight
  exXw:  exam weight
  hXa:  homework average
  exXa:  exam average
-->

<!--- count the students -->

<xsl:variable name="ns">
  <xsl:value-of select="count(student)"/>
</xsl:variable>

<!--- homework and exam max -->

<xsl:variable name="hw1m">
  <xsl:value-of select="hw1/@max"/>
</xsl:variable>
```

```
<xsl:variable name="hw2m">
  <xsl:value-of select="hw2/@max"/>
</xsl:variable>

  <xsl:variable name="ex1m">
  <xsl:value-of select="ex1/@max"/>
</xsl:variable>

<xsl:variable name="ex2m">
  <xsl:value-of select="ex2/@max"/>
</xsl:variable>

<!--- homework and exam weights -->

<xsl:variable name="hw1w">
  <xsl:value-of select="hw1/@weight"/>
</xsl:variable>

<xsl:variable name="hw2w">
  <xsl:value-of select="hw2/@weight"/>
</xsl:variable>

<xsl:variable name="ex1w">
  <xsl:value-of select="ex1/@weight"/>
</xsl:variable>

<xsl:variable name="ex2w">
  <xsl:value-of select="ex2/@weight"/>
</xsl:variable>

<!--- homework and exam normalization factor-->

<xsl:variable name="norm">
  <xsl:value-of select="$hw1m*$hw1w + $hw2m*$hw2w
        +$ex1m*$ex1w + $ex2m*$ex2w" />
</xsl:variable>

<!--- homework and exam averages-->

<xsl:variable name="hw1a">
  <xsl:value-of select="sum(student/hw1/@grade) div $ns"/>
</xsl:variable>

<xsl:variable name="hw2a">
  <xsl:value-of select="sum(student/hw2/@grade) div $ns"/>
</xsl:variable>

<xsl:variable name="ex1a">
  <xsl:value-of select="sum(student/ex1/@grade) div $ns"/>
```

```
</xsl:variable>

<xsl:variable name="ex2a">
  <xsl:value-of select="sum(student/ex2/@grade) div $ns"/>
</xsl:variable>

<!-- generate html code -->

<head>
  <title>
    <xsl:value-of select="@name"/> Grades
  </title>
</head>

<!-- html styles -->

<style type="text/css">
table.grades{
  border-width: 1px;
  border-spacing: 2px;
  border-style: outset;
  border-color: gray;
  border-collapse: separate;
  background-color: wheat;
  font-size:12px;
}
table.grades th {
  border-width: 1px;
  padding: 1px;
  border-style: inset;
  border-color: gray;
  background-color: white;
  color: maroon;
  font-size:14px;
}
table.grades td {
  border-width: 1px;
  padding: 1px;
  border-style: inset;
  border-color: gray;
}
</style>

<body>

<!-- print the course name and term -->

<xsl:value-of select="@name"/>,
<xsl:value-of select="@term"/> <br/><br/>
```

```
<!-- table -->

<table class="grades">

<tr>
<th>Student</th><th>Hw 1</th><th>Hw 2</th><th>Ex 1</th>
    <th>Ex 2</th> <th>Score</th> <th>Grade</th>
</tr>

<xsl:for-each select ="student">
  <xsl:sort select="firstname" order="ascending"/>
  <tr>

  <td align="left">
    <xsl:value-of select="lastname"/> &#173;
    <xsl:value-of select="firstname"/> &#173;
    <xsl:value-of select="middlename"/>
  </td>
  <td align="right"><xsl:value-of select="hw1/@grade"/></td>
  <td align="right"><xsl:value-of select="hw2/@grade"/></td>
  <td align="right"><xsl:value-of select="ex1/@grade"/></td>
  <td align="right"><xsl:value-of select="ex2/@grade"/></td>
  <td align="right">
    <xsl:variable name="grdn">
      <xsl:value-of select="(( hw1/@grade * $hw1w + hw2/@grade * $hw2w
      + ex1/@grade * $ex1w + ex2/@grade * $ex2w ) * 100 ) div $norm"/>
    </xsl:variable>
    <xsl:value-of select='format-number($grdn,"##.##")'/>
  </td>

  <td align="center">
    <xsl:variable name="grdn">
      <xsl:value-of
        select=" ((hw1/@grade * $hw1w + hw2/@grade * $hw2w
        + ex1/@grade * $ex1w + ex2/@grade * $ex2w )
        * 100 ) div $norm"/>
    </xsl:variable>

    <xsl:choose>
      <xsl:when test="($grdn &gt; 95) and ($grdn &lt; 101)">
        A</xsl:when>
      <xsl:when test="($grdn &gt; 90) and ($grdn &lt; 95)">
        B</xsl:when>
      <xsl:when test="($grdn &gt; 85) and ($grdn &lt; 90)">
        C</xsl:when>
      <xsl:when test="($grdn &gt; 80) and ($grdn &lt; 85)">
        D</xsl:when>
      <xsl:otherwise>F</xsl:otherwise>
```

```
    </xsl:choose>
  </td>

</tr>
</xsl:for-each>

<!-- maximum -->

<tr bgcolor="ghostwhite">
  <td align="center">Maximum</td>
  <td align="right"><xsl:value-of select="hw1m"/></td>
  <td align="right"><xsl:value-of select="hw2m"/></td>
  <td align="right"><xsl:value-of select="ex1m"/></td>
  <td align="right"><xsl:value-of select="ex2m"/></td>
  <td align="right"></td>
  <td align="right"></td>
</tr>

<!-- averages -->

<tr bgcolor="ghostwhite">
<td align="center">Average </td>
<td align="right"><xsl:value-of
    select='format-number($hw1a,"##.##")'/></td>
<td align="right"><xsl:value-of
    select='format-number($hw2a,"##.##")'/></td>
<td align="right"><xsl:value-of
    select='format-number($ex1a,"##.##")'/></td>
<td align="right"><xsl:value-of
    select='format-number($ex2a,"##.##")'/></td>
<td align="right"></td>
<td align="right"></td>
</tr>

<!-- end of a table -->

</table> <br/>

<xsl:value-of
qquad select="count(student)"/> students in this class <br/><br/>

</body>

</html>
</xsl:template>
</xsl:stylesheet>
```

The *xsl* implementation makes extensive use of variables to compute averages. The *sort xslt* element is placed inside the *for-each* loop to print the student records in alphabetical order. The space character () is printed to separate the last, from the first, from the middle name of each student in the output.

Opening the *eigenvalues.xml* file with a *web* browser produces the following display:

```
ENG101, Spring 2005

Student                 Hw 1   Hw 2    Ex 1    Ex 2    Score   Grade
Hildebrand Anne R.       92     83      57      70     94.85     B
Smith Jim R.             80     70      30      30     54.55     F
Sotiris Kathryne L.      90     59      55      55     80.91     D
Maximum                 100    100      60      70
Average                  87.33  70.67   47.33   51.67

3 students in this class
```

This output lacks nothing in style and content compared to that produced by a professional spreadsheet. Additional features can be easily incorporated.

Exercise

4.9.1 *Final exam*

Add to the student roster the scores of the final exam.

4.9.2 *Maximum grade*

Add to the student roster a row printing the maximum grade earned by the students in each homework and each exam.

5 *Producing and importing xml data*

Numbers generated by a scientific code can be recorded in a file according to *xml* conventions and grammar. These *xml* data may then be imported for processing into another code or viewed in a *web* browser with the aid of an *xsl* stylesheet, as discussed in Chapters 3 and 4. Three main tasks of interest in scientific computing are (*a*) creating *xml* formatted output from scientific code, (*b*) reading *xml* formatted input and accommodating it into data structure of a chosen programming language, and (*c*) navigating through an *xml* document with the objective of retrieving desired pieces of information.

The simplest method of generating *xml* output in any computer language is by hard-coding *print* statements for the individual *xml* elements, including tags, attributes, and element content. Although this approach could be belittled, scientific programmers are known to favor convenient shortcuts that ameliorate the steepness of a learning curve and minimize the net programming effort. Hard-coding of *xml* data is used routinely for sending formatted information packets over the Internet.

Importing data from an *xml* document into data structures of a chosen computer language can be cumbersome, especially when files of large size are involved. In one shortcut, *xml* tags are stripped to produce data files containing columns using a suitable string manipulation language, such as *perl*, *sed*, or *awk*. The unformatted files are used as input, subject to implied conventions. In a related shortcut, when running a code, an *xml* formatted file and another unformatted text file containing the same data are produced in the output. The former is used for *xml* documentation and the latter is used as input to the same or another scientific code.

Language libraries and modules that generate and import *xml* data are available in several scientific languages that support advanced data structures and object-oriented programming, such as C++, *perl*, and *java*. Only a limited number of such facilities are available in procedural languages, such as *fortran*. *Perl* is particularly attractive due to the availability of associative arrays (hashes) that can be used to define and describe in words heterogeneous objects, as discussed in Section 1.6.4 and Appendix B. Scientific computing languages

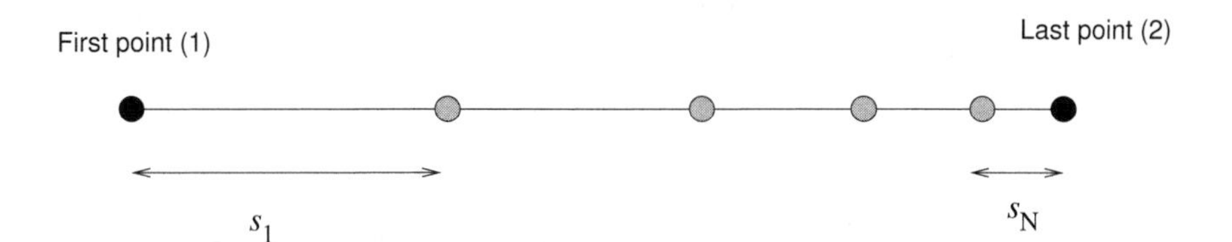

FIGURE 5.1 Discretization of a straight segment into N elements. The length of the elements, s, increases or decreases geometrically from the first segment point to the last segment point so that $s_N/s_1 =$ `ratio` has a specified value. In the illustration shown, `ratio` < 1.

that take advantage of *xml* data structures in the spirit of *xsl* are not available at the present time.

In this chapter, possible ways of handling *xml* data in *fortran*, *perl*, C++, and *Matlab* are discussed with the objective of illustrating typical procedures.

5.1 Fortran

Fortran 77 is a remarkable computer language with a long and distinguished history in scientific computing. *Xml* data can be produced by a *fortran* code using formatted *print* or *write* statements. The procedure will be illustrated in this section by an example.

Discretization of a straight segment

We are interested in dividing a straight segment in the xy plane, beginning at a point, (x_1, y_1), and ending at another point, (x_2, y_2), into N straight elements. The length of the elements should increase or decrease geometrically by a constant factor, α, as shown in Figure 5.1. This means that

$$s_{m+1} = \alpha s_m = \alpha^2 s_{m-1} = \cdots = \alpha^m s_1 \tag{5.1}$$

for $m = 1, \ldots, N-1$, where s_i is the length of the ith element. Consequently,

$$r \equiv \frac{s_N}{s_1} = \alpha^{N-1} \tag{5.2}$$

and

$$\alpha = r^{1/(N-1)}, \tag{5.3}$$

where r is a specified ratio. The total length of the line is

$$L = s_1 + s_2 + \cdots + s_N = s_1\,(1 + \alpha + \cdots + \alpha^{N-1}) = s_1 \frac{1-\alpha^N}{1-\alpha}. \tag{5.4}$$

Inverting this relation, we obtain the length of the first segment in terms of the total length of the line,

$$s_1 = \frac{1-\alpha}{1-\alpha^N}\, L. \tag{5.5}$$

Provisions must be made for the special case $N = 1$ or when $r = 1$ where all elements have the same length, $\alpha = 1$ and $s_1 = L/N$.

Discretization subroutine

The discretization is performed by a subroutine contained in the following file entitled *elm_line1.f* where the meaning of the variables employed is described in a legend:

```
      subroutine elm_line1
     +
     + (N
     + ,ratio
     + ,X1,Y1
     + ,X2,Y2
     + ,sinit
     + ,Xe,Ye,se
     + ,Xm,Ym,sm
     + )

c-----------------------------------------------------
c Discretization of a line segment
c into a graded mesh of N elements
c
c X1,Y1:  coordinates of the first point
c X2,Y2:  coordinates of the last point
c ratio:  element stretch ratio
c alpha:  geometric factor ratio
c sinit:  specified arc length at (X1, Y1)
c
c se:  arc length at the element end-nodes
c sm:  arc length at the element mid-nodes
c
c xe,ye:  element end nodes
c xm,ym:  element mid nodes
c-----------------------------------------------------

      Implicit Double Precision (a-h,o-z)

      Dimension Xe(129),Ye(129),se(129)
      Dimension Xm(128),Ym(128),sm(128)
```

```
c------------
c one element
c------------

      If(N.eq.1) then
        xe(1) = X1
        ye(1) = Y1
        xe(2) = X2
        ye(2) = Y2
        se(1) = sinit
        se(2) = se(1)+Dsqrt((X2-X1)**2+(Y2-Y1)**2)
        Go to 99
      End If

c--------------
c many elements
c--------------

      If(ratio.eq.1.000) then
        alpha = 1.0D0
        factor = 1.0D0/N
      Else
        texp = 1.0D0/(N-1.0D0)
        alpha = ratio**texp
        factor = (1.0D0-alpha)/(1.0D0-alpha**N)
      End If

      write (6,*) ratio
      write (6,*) "elm_line:  Geometric ratio:  ",alpha

      deltax = (x2-x1) * factor !  x length of first element
      deltay = (y2-y1) * factor !  y length of first element

      Xe(1) = X1 !  first point
      Ye(1) = Y1
      se(1) = sinit

      Do i=2,N+1
        Xe(i) = Xe(i-1)+deltax
        Ye(i) = Ye(i-1)+deltaY
        se(i) = se(i-1)+sqrt(deltax**2+deltay**2)
        deltax = deltax*alpha
        deltay = deltay*alpha
      End Do

c-----
c Done
c-----
```

```
  99  Continue

c------------------
c compute mid-points
c------------------

      Do i=1,N
        xm(i) = 0.5D0*(xe(i)+xe(i+1))
        ym(i) = 0.5D0*(ye(i)+ye(i+1))
        sm(i) = 0.5D0*(se(i)+se(i+1))
      End Do

c-----
c Done
c-----

      Return
      End
```

Note the presence of six obligatory blank spaces at the beginning of each line. The `+` sign in the sixth line serves as a continuation mark. The comment mark character `c` appears at the beginning of each line where explanations and commentary are provided. The arguments of the subroutine enclosed by parentheses,

```
elm_line1( ... )
```

include both the input and the output. In contrast, in C, C++, and *Matlab*, the input is separated from the output in function calls.

The discretization subroutine is called by the main program *elm_line1_dr.f* residing in a driver file listed below:

```
      program elm_line1_dr

c---------------------------------------
c Element discretization of a line
c---------------------------------------

      Implicit Double Precision (a-h,o-z)
      Dimension xe(129),ye(129)

c-----
c data
c-----

      X1 = 0.0D0
      Y1 = 0.0D0
      X2 = 1.0D0
```

```
      Y2 = 0.4D0
      N = 8
      ratio =2.0D0

      call elm_line1
     +
     + (N
     + ,ratio
     + ,X1,Y1
     + ,X2,Y2
     + ,Xe,Ye
     + )

c-----
c print xml
c-----

      open (1,file="elm_line1.xml")
      write (1,100)
      write (1,101)
      write (1,*) "<grid>"
      write (1,*)
      Do i=1,N
        write (1,*) ' <element id="',i,'">'
        write (1,*) ' <node1>'
        write (1,201) xe(i)
        write (1,202) ye(i)
        write (1,*) ' </node1>'
        write (1,*) ' <node2>'
        write (1,201) xe(i+1)
        write (1,202) ye(i+1)
        write (1,*) ' </node2>'
        write (1,*) " </element>"
      End Do
      write (1,*)
      write (1,*) "</grid>"
      close(1)

c-----
c Done
c-----

100 Format('<?xml version="1.0" encoding="ISO-8859-1"?>')
101 Format('<?xml-stylesheet type="text/xsl"href="elm_line1.xsl"?>')
201 Format (10X,'<x>',f10.5,'</x>')
202 Format (10X,'<y>',f10.5,'</y>')

Stop
End
```

A file labeled 1 and named *elm_line1.xml* is opened by the statement:

```
open (1,file="elm_line1.xml")
```

and closed near the end of the code by the statement:

```
close (1)
```

The statement `write (1,*)` prints in this file specified data with a free format chosen by the compiler, indicated by the asterisk. The statement `write (1,101)` prints specified data with the format statement numbered `101` located near the end of the code (the exact placement is arbitrary). Other formatting statements are employed in the code.

Compilation and execution

The main program and subroutine can be compiled by opening a terminal (command-line window) and issuing the statement:

```
f77 -c elm_line1_dr.f
```

followed by the ENTER keystroke, and then the statement:

```
f77 -c elm_line1.f
```

followed by the ENTER keystroke. Two object files generated in this fashion, named *elm_line1.o* and *elm_line1_dr.o*, can be linked into an executable binary file named *elm_line1* by issuing the statement:

```
f77 -o elm_line1 elm_line1.o elm_line1_dr.o
```

and then pressing the ENTER key.

For convenience, the executable can be generated following the instructions contained in the following *makefile* defining dependencies and procedures:

```
elm_line1:  elm_line1.o elm_line1_dr.o
    f77 -o elm_line1 elm_line1.o elm_line1_dr.o
elm_line1_dr.o:  elm_line1_dr.f
    f77 -c elm_line1_dr.f
elm_line1.o:  elm_line1.f
    f77 -c elm_line1.f
```

where the indentation is produced by pressing the TAB key. A makefile provides input to the *make* program (application) that receives and executes multiple sets of instructions. To create the executable, we open a *unix* terminal and issue the statement:

```
make elm_line1
```

Once we have created the executable, we can run it by typing:

```
./elm_line1
```

and then pressing the ENTER key. The code generates the following file named *elm_line1.xml*:

```
<?xml version="1.0" encoding="ISO-8859-1"?>
<?xml-stylesheet type="text/xsl"href="elm_line1.xsl"?>
<grid>

  <element id="1">
    <node1>
      <x>0.00000</x>
      <y>0.00000</y>
    </node1>
    <node2>
      <x>0.08615</x>
      <y>0.03446</y>
    </node2>
  </element>
    ...
  <element id="8">
    <node1>
      <x>0.82769</x>
      <y>0.33108</y>
    </node1>
    <node2>
      <x>1.00000</x>
      <y>0.40000</y>
    </node2>
  </element>

</grid>
```

where the three dots represent additional lines of data describing physical straight elements arising from the discretization. The name of the root element in this *xml* document is `grid`.

Xsl processing of the xml output

Next, we write an *xsl* file entitled *elm_line1.xsl* to process the *xml* data and display the results in a *web* browser. The content of this file is:

```
<xsl:stylesheet version="1.0"
    xmlns:xsl="http://www.w3.org/1999/XSL/Transform">
<xsl:template match="grid">
<html>
```

```
<body>
<center>

  <h2>Boundary element discretization</h2>

  <table border="2" cellspacing="4" cellpadding="4">
    <tr bgcolor="olive">
      <th>Element</th>
      <th colspan="2">First Node</th>
      <th colspan="2">Second Node</th></tr>
    <tr bgcolor="wheat">
      <th></th> <th>x</th> <th>y</th> <th>x</th> <th>y</th>
    </tr>
    <xsl:for-each select="grid/element">
      <tr bgcolor="ghostwhite" align="center">
      <td><xsl:value-of select="@id"/></td>
      <td><xsl:value-of select="node1/x"/></td>
      <td><xsl:value-of select="node1/y"/></td>
      <td><xsl:value-of select="node2/x"/></td>
      <td><xsl:value-of select="node2/y"/></td>
      </tr>
    </xsl:for-each>

  </table>

</center>
</body>

</html>
</xsl:template>
</xsl:stylesheet>
```

The *xml* file can be manipulated using the *xsltproc* processor. Opening a terminal (command-line window) and issuing the statement:

```
xsltproc elm_line1.xml
```

followed by the ENTER keystroke, prints on the screen the following *html* code:

```
<html><body><center>
<h2>Boundary element discretization</h2>
<table border="2" cellspacing="4" cellpadding="4">
<tr bgcolor="olive">
<th>Element</th>
<th colspan="2">First Node</th>
<th colspan="2">Second Node</th>
</tr>
<tr bgcolor="wheat">
<th></th>
```

Boundary element discretization

Element	First Node		Second Node	
	x	y	x	y
1	0.00000	0.00000	0.08615	0.03446
2	0.08615	0.03446	0.18128	0.07251
3	0.18128	0.07251	0.28630	0.11452
4	0.28630	0.11452	0.40225	0.16090
5	0.40225	0.16090	0.53028	0.21211
6	0.53028	0.21211	0.67163	0.26865
7	0.67163	0.26865	0.82769	0.33108
8	0.82769	0.33108	1.00000	0.40000

FIGURE 5.2 Data displayed in a *web* browser based on an *xml* document generated by a *fortran* code and processed by a companion *xsl* code.

```
...
</table>
</center></body></html>
```

where the three dots indicate additional lines. Alternatively, we may open the *elm_line1.xml* file with a *web* browser to obtain the display presented in Figure 5.2.

The example discussed in this section illustrates the simplicity and ease of implementation of *fortran*. Legacy code written in *fortran* is used today in numerous science and engineering applications.

Importing xml data

Writing *fortran* subroutines that read data from an *xml* file with an arbitrary tree structure can be daunting. Because *fortran* is meant to be used for numerical computation, it is not suitable for character and string manipulation. The opposite for the *xsl* language discussed in Chapters 3 and 4, and for other system programming or object-oriented languages, such as *perl* and C++, as discussed later in this chapter. Our best option for importing *xml* data into a *fortran* code is to employ available *xml* libraries containing suitable import and export functions. Alternatively, we may use *fortran* for numerical computation and another language for *xml* data manipulation.

Fortran/xml libraries

A small number of *fortran* subroutines and libraries capable of handling *xml* data have been written in *fortran* 95.[*][†][‡] *Fortran* 95 and its precursor *fortran* 90 differ significantly from *fortran* 77, to the extent that they could be regarded as different languages. These *xml* libraries allow us to read and write *xml* files from *fortran* code.

Exercises

5.1.1 *Fortran code*

Write a small *fortran* code of your choice.

5.1.2 *Hard-coded xml*

Write a *fortran* code that produces *xml* formatted data of your choice in the output. Then write a companion *xsl* file that displays the data in the window of a *web* browser.

5.1.3 *Fortran/xml libraries*

Discuss the capabilities and prepare an overview of a *fortran/xml* library of your choice.

5.2 Perl

The *perl* programming language was introduced in Section 1.6.4 with special attention to *perl* hashes. A brief outline of the basic language structure can be found in Appendix B. *Perl* employs scalar variables, homogeneous arrays, and inhomogeneous arrays that may contain the same or different data types, including integers, real numbers, and character strings.

A *perl* hash is an associative array linking keys (variable names) to values (numbers or strings). Thus, a *perl* hash defines and describes data in terms of keys. To define or extract the content of a *perl* hash, we prepend the percent sign (%) to the hash name.

Perl employs modules (*pm*) playing the role of language libraries. *Xml* modules are available, thanks to independent developers. For example, a *perl* module implementing an interface to the comprehensive library *libxml* is available.[§] *Libxml* is an *xml* parser and toolkit written in C, developed for the *gnome* project.

[*] http://xml-fortran.sourceforge.net
[†] http://nn-online.org/code/xml
[‡] http://fortranwiki.org/fortran/show/FoX
[§] http://cpan.uwinnipeg.ca/dist/XML-LibXML

5.2.1 XML::Simple

We will discuss an elegant *xml* module named *XML::Simple* written by Grant McLean.* The current version of *XML::Simple* numbered 2.18 was released in August, 2007.† A simpler version of *XML::Simple*, named *XML::Simpler*, was released in April 2002.‡ In this section, we will discuss the original version.

The functions implemented in *XML::Simple* allow us to extract data from an *xml* file, placing them into a *perl* data structure. Specifically, *XML::Simple* parses an *xml* document and maps an element tree onto an anonymous *perl* hash accessed by a reference. The *xml* elements may then be retrieved using standard *perl* array manipulations. Conversely, *XML::Simple* allows us to print *xml* data from native *perl* data structures consistent with an *xml* tree.

XML::Simple employs the *XML::Parser perl* module. Both modules are incorporated in most *perl* distributions. Alternatively, *XML::Simple* can be downloaded from the comprehensive perl archive network (*cpan*) using the *cpan* shell.

Installation

To install *XML::Simple* in a *unix* system, we open a terminal (command-line window) and issue the command:

```
$ sudo perl -MCPAN -e shell
```

followed by the ENTER keystroke, where the dollar sign ($) is a *unix* shell prompt. After entering the administrative password, we enter the *cpan* environment and issue the command:

```
cpan[1]> install XML::Simple
```

where `cpan[1]>` is a *cpan* prompt. When the installation finishes, we exit by typing:

```
cpan[2]> quit
```

Dependencies are automatically downloaded.

In *Windows*, we open a terminal (command-line window) and issue the command:

```
$ ppm install XML::Simple
```

*http://www.mclean.net.nz/cpan
†http://search.cpan.org/ grantm/XML-Simple
‡http://www.mclean.net.nz/cpan/xmlsimpler/XML-Simpler.html

Summary of XML::Simple functions

To use *XML::Simple*, we insert the following statement at the beginning of a *perl* code declaring the use of the *XML::Simple* module:

```
use XML::Simple;
```

XML::Simple allows us to call two main functions for reading and writing *xml* data.

The function

```
$somereference = XMLin("somefile.xml", someoptions);
```

reads an *xml* file and returns a reference to an anonymous hash. By convention, *XMLin* reads an *xml* file called *somename.xml*, whose name is the same as that of the *perl* file calling the function, *somename.pl*.

The function

```
$xmlreference = XMLout($hashreference, options);
```

converts a hash represented by a reference to an *xml* document represented by another reference.

Constructing and evaluating objects

Alternatively, an *XML::Simple* object accessible by a reference can be constructed by issuing the statement:

```
$somereference = XML::Simple->new(options);
```

or the statement:

```
$somereference = new XML::Simple(options);
```

The object can be subsequently evaluated by issuing the statement:

```
$otherreference = $somereference->XMLin("somefile.xml",options);
```

where `otherreference` is a reference to the evaluated object.

To convert a hash reference into an *xml* document represented by another reference, we use the line:

```
$xmlreference = $somereference->XMLout($hashreference, options);
```

5.2.2 Roots of an equation

As an example, we consider the following data recorded in a file entitled *roots.xml* representing the real and imaginary parts of the roots (zeros) of an equation:

```
<?xml version="1.0" encoding="ISO-8859-1"?>
<rizes>

<root>
  <real>-3.78985</real>
  <imaginary>0.00000</imaginary>
</root>

<root>
  <real>2.89492</real>
  <imaginary>3.09226</imaginary>
</root>

<root>
  <real>28.9872</real>
  <imaginary>-1.18236</imaginary>
</root>

</rizes>
```

The name of the root element is `rizes`.

The following *perl* script residing in a file entitled *roots.pl* parses the *xml* document using the function *XMLin* of the *XML::Simple* module:

```
#!/usr/bin/perl
#
use XML::Simple;              # use a perl module
use Data::Dumper;             # use a perl module
#
$gamgee = XMLin("roots.xml", KeepRoot=>1);    # read the xml file
#
print Dumper($gamgee);      # print the hash
```

The individual lines of this script perform the following functions:

- The first statement reveals the directory where the *perl* interpreter resides in our *unix* operating system.
- The second line is an empty comment inserted for visual clarity.
- The third and fourth lines include two *perl* modules.
- The fifth line is an empty comment inserted for visual clarity.

- The *XMLin* function is used in the sixth line to read the *xml* file and store the data in an anonymous hash structure referenced by the variable `gamgee`.
- The seventh line is an empty comment inserted for visual clarity.
- The eighth line requests printing the *perl* data structure referenced by `gamgee` using the *Data::Dumper* module.*

The `KeepRoot` option inside the arguments of the *XMLin* function is a Boolean flag (yes or no) used to discard or preserve the root element of the *xml* document.

Running the script

To run the *perl* script, we open a terminal (command-line window) and type the name of the script

```
./roots.pl
```

followed by the ENTER keystroke. To ensure that the path of executables includes the current directory, we have inserted a dot-slash pair (`./`), indicating the current directory, in front of the *perl* file name. The generated display is shown in Table 5.1.

We observe that the *Data::Dumper* module generates a reference named `VAR1` to an anonymous *perl* hash referenced by the scalar `gamgee`, enclosed by the outer curly brackets (`{}`). The anonymous hash has a single key named `rizes`, which is the name of the root element of the *xml* file. This key represents an anonymous array of anonymous hashes enclosed by the innermost curly brackets (`{}`). The elements of the anonymous array are enclosed by square brackets (`[]`).

Scientific computing programmers will interpret the *perl* structure referenced by `gamgee` as a generalized vector whose elements are enclosed by the outer curly brackets (`{}`). The square brackets (`[]`) are containers of an anonymous vector. The innermost curly brackets hold additional vectors (associative arrays). Overall, the generated *perl* structure consists of a tree-like cascade of vectors reminiscent of an *xml* tree.

The last three lines of the script could have been consolidated into the single line:

```
print Dumper( XML::Simple->new->XMLin("roots.xml", KeepRoot=>1)) );
```

However, with this choice, a reference to a *perl* object will not be created.

---

*http://search.cpan.org/ smueller/Data-Dumper-2.131

```
$VAR1 = {
          'rizes' => {
                       'root' => [
                                   {
                                     'imaginary' => ' 0.00000',
                                     'real' => '-3.78985'
                                   },
                                   {
                                     'imaginary' => ' 3.09226',
                                     'real' => ' 2.89492'
                                   },
                                   {
                                     'imaginary' => '-1.18236',
                                     'real' => ' 28.9872'
                                   }
                                 ]
                     }
        };
```

TABLE 5.1 An anonymous *perl* hash accommodating specified *xml* data generated by the *XMLin* function of *XML::Simple*. The anonymous hash is referenced by the scalar VAR1 containing a single key named root representing an anonymous array of anonymous hashes.

Simple objects

The important line:

```
$gamgee = XMLin("roots.xml", KeepRoot=>1);
```

generating a reference to a *perl* structure, could have been replaced by the lines:

```
$peregrine = XML::Simple->new();       # create an object
$gamgee = $peregrine->XMLin("roots.xml");      # evaluate the object
```

The first line defines a reference to a new *XML::Simple* object and the second line defines a reference to the evaluated object.

Retrieving data

To print the real and imaginary parts of the first root, we insert the following statements in the *perl* script:

```
print $gamzee->{rizes}{root}[0]{real},"\n";
print $gamzee->{rizes}{root}[0]{imaginary},"\n";
```

The dereference operator denoted by an *ascii* arrow (`->`) leads us from an object reference, in this case `gamgee`, to the object itself. Two keys, `rizes` and `root`, are enclosed by curly brackets (`{}`). Anonymous array indices, 0 and 1, are enclosed by square brackets (`[]`). Two inner keys, `real` and `imaginary`, are enclosed by curly brackets (`{}`). Printing the `\n` character forces a line feed. The display generated by these lines is:

```
-3.78985
0.00000
```

To print the real and imaginary parts of all roots, we append the following lines to the *perl* script:

```
foreach $item (0,1,2)
{
print "Root ",$item+1," :"
     ," Real:  ",$gamzee->{rizes}{root}[$item]{real}
     ," Imag:  ",$gamzee->{rizes}{root}[$item]{imaginary}
     ,"\n";
}
```

The generated screen display is:

```
Root 1: Real: -3.78985 Imag: 0.00000
Root 2: Real:  2.89492 Imag: 3.09226
Root 3: Real: 28.9872 Imag: -1.18236
```

The `foreach` loop can be replaced with a more general loop handling an arbitrary number of roots:

```
foreach $ind (@{$gamzee->{rizes}{root}})
{
print "Real:  ",$ind->{real}," Imag:  ",$r->{imaginary},"\n";
}
```

The variable `ind` runs over an array consisting of the components of the `root` array. The produced screen display is:

```
Real: -3.78985 Imag: 0.00000
Real: 2.89492 Imag: 3.09226
Real: 28.9872 Imag: -1.18236
```

Printing to a file

To print the data in a file named *roots.txt*, we open the file under the arbitrary alias `barney`, print the data, and then close the file according to the lines:

```
open(barney, '> roots.txt');

foreach $r (@$gamzee->{rizes}{root})
{
   print barney "Pragmatikos: ",$r->{real},
    " Fantastikos: ",$r->{imaginary},"\n";
}

close(barney)
```

The content of the file *roots.txt* generated by this code is:

```
Pragmatikos: -3.78985 Fantastikos: 0.00000
Pragmatikos: 2.89492 Fantastikos: 3.09226
Pragmatikos: 28.9872 Fantastikos: -1.18236
```

Because *barney* was opened with a single pointer (>) instead of two pointers (>>), the data printed in the *roots.txt* file replaces the previous content of *roots.txt*, if any.

5.2.3 Real, imaginary, and complex roots

Now we consider data contained in an *xml* file entitled *primm.xml* where real roots (rroot), imaginary roots (iroot), and complex roots (croot) of an equation are separately recorded. The content of this file is:

```
<?xml version="1.0" encoding="ISO-8859-1"?>
<roots>

  <rroot>-3.0</rroot>
  <rroot>2.5</rroot>
  <iroot>1.0</iroot>
  <iroot>2.00</iroot>

  <croot>
    <real>28.00</real> <imaginary>-1.3</imaginary>
  </croot>

  <croot>
    <real>14.00</real> <imaginary>9.2</imaginary>
  </croot>

</roots>
```

The name of the root element is `roots`. The following *perl* script named *primm.pl* parses this file using the `XML::Simple` module and returns a reference named `gamgee` to an anonymous *perl* hash:

```
#!/usr/bin/perl
#
use XML::Simple;      # use a perl module
use Data::Dumper;      # use a perl module
#
$gamgee = XMLin("primm.xml",KeepRoot=>1);      # read the xml file
print Dumper($gamgee);      # print
```

Running this script generates the display shown in Table 5.2(*a*).

We see that a reference (`VAR1`) to an anonymous hash referenced by `gamgee` appears. The anonymous hash has a single key named `roots` pointing to an interior anonymous hash containing three keys named `rroot`, `croot`, and `irroot`. Each one of these keys points to an anonymous array of scalars or anonymous hashes.

The corresponding data tree is shown in Table 5.2(*b*). The numbers inside the square brackets are indices of arrays, starting at zero, representing a node list.

Retrieving data

Appending to the script the lines:

```
print $gamgee->{roots}{rroot}[0], "\n ";
print $gamgee->{roots}{rroot}[1], "\n";
print $gamgee->{roots}{iroot}[0], "\n  ";
print $gamgee->{roots}{iroot}[1], "\n";
print $gamgee->{roots}{croot}{real}, "  ";
print $gamgee->{roots}{croot}{imaginary}, "\n";
```

produces the screen display:

```
-3.0
 2.5
 1.0
 2.00
28.0     -1.3
```

These statements illustrate how different data types can be accessed and manipulated in a *perl* code.

5.2.4 Molecules

In the next example, we consider an *xml* document whose elements contain mixed data types, including numbers and character strings.

(*a*)

```
$VAR1 = {
          'roots' => {

                       'rroot' => [
                                    '-3.0',
                                    ' 2.5'
                                  ],
                      'croot' => [
                                   {
                                    'imaginary' => '-1.3',
                                    'real' => ' 28.0'
                                   },
                                   {
                                    'imaginary' => '9.2',
                                    'real' => '14.0'
                                   }
                                   ],
                       'iroot' => [
                                    '1.0',
                                    '2.00'
                                    ]
                       }
        };
```

(*b*)

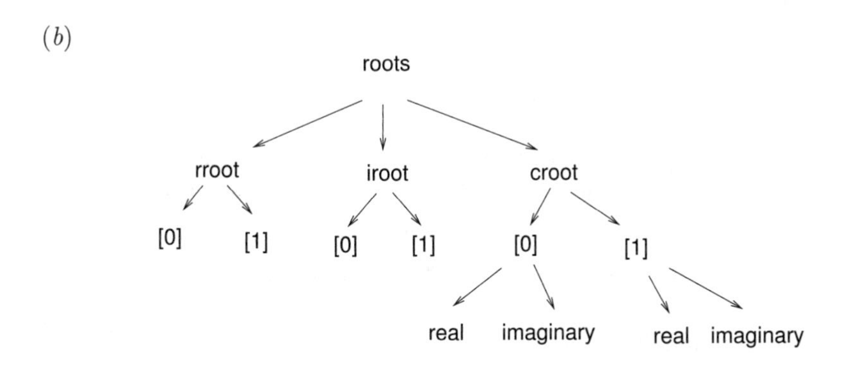

TABLE 5.2 (*a*) An anonymous *perl* hash referenced by the scalar VAR1 generated by the *XMLin* function of *XML::Simple*. (*b*) The corresponding data tree. The indices enclosed by square brackets represent elements of an anonymous array.

The following *xml* file entitled *particles.xml* describes two molecules and one atom:

```
<?xml version="1.0" encoding="ISO-8859-1"?>
<somatidia>

   <molecule>
      <name>water</name>
      <atoms>3</atoms>
   </molecule>

   <molecule>
      <name>methanol</name>
      <atoms>6</atoms>
   </molecule>

   <atom>
      <name>carbon</name>
   </atom>

</somatidia>
```

The name of the root element is `somatidia`. The following *perl* script named *particles.pl* processes this file using the `XML::Simple` module:

```
#!/usr/bin/perl
#
use XML::Simple; # use a perl module
use Data::Dumper; # use a perl module
#
$arwin = XMLin("particles.xml"); # reference to a perl structure
print Dumper($arwin); # print
```

Running the script generates the display shown in Table 5.3(a). For brevity, the root element of the *xml* file has been discarded in the *perl* hash array. Issuing the statement:

```
print $arwin->{molecule}{methanol}{atoms},"\n";
```

prints the expected number of atoms,

```
6
```

The chain leading us to the target involves three nested anonymous *perl* hashes.

(*a*)

```
$VAR1 = {
          'atom' => {
                      'name' => 'carbon'
                    },
          'molecule' => {
                          'methanol' => {
                                          'atoms' => '6'
                                        },
                          'water' => {
                                       'atoms' => '3'
                                     }
                        }
        };
```

(*b*)

```
$VAR1 = {
          'atom' => {
                      'name' => 'carbon'
                      },
          'molecule' => [
                           {
                             'name' => 'water',
                             'atoms' => '3'
                           },
                           {
                             'name' => 'methanol',
                             'atoms' => '6'
                           }
                         ]
        };
```

TABLE 5.3 An anonymous *perl* hash referenced by the scalar VAR1 generated by the *XMLin* function of *XML::Simple* (*a*) without and (*b*) with the *key attribute option*.

Key attribute

The data tree encapsulated in the *perl* structure shown in Table 5.3(*a*) is not entirely satisfactory. We much prefer that the name and number of atoms of each molecule appear together under the same hash with different keys. To achieve this goal, we generate the *perl* structure of the *xml* data using the `KeyAttr` option, by issuing the statement:

```
$arwin = XMLin("molecules.xml", KeyAttr=>[]);
```

Running the script produces an anonymous hash with the desired structure, as shown in Table 5.3(*b*).

Next, we append to the *perl* script the following lines:

```
foreach  $ind (@{$arwin->{molecule}})
{
print "The ",$ind->{name}," molecule has ",$ind->{atoms}," atoms\n";
}
```

The screen display generated by these lines is:

```
The water molecule has 3 atoms
The methanol molecule has 6 atoms
```

The complete *perl* script can be found in a file entitled *prtcl.pl.*

Force Array

As an experiment, we create a *perl* object using the line:

```
$gimli = XMLin("molecules.xml", ForceArray => 1);
```

The ForceArray option is a Boolean flag (yes or no) used to store *xml* elements into regular indexed arrays instead of hashes. The modified script resides in a file entitled *parted.pl.* Running the script prints on the screen the structure shown in Table 5.4. Appending to the script the lines:

```
foreach $mol (@{$nevada->{molecule}})
  {
    print "The ",$mol->{name}[0]," molecule has ",
              $mol->{atoms}[0]," atoms\n";
  }
```

prints on the screen the following lines:

```
$VAR1 = {
          'atom' => [
                      {
                        'name' => [
                                    'carbon'
                                  ]
                      }
                    ],
          'molecule' => [
                          {
                            'name' => [
                                        'water'
                                      ],
                            'atoms' => [
                                         '3'
                                       ]
                          },
                          {
                            'name' => [
                                        'methanol'
                                      ],
                            'atoms' => [
                                         '6'
                                       ]
                          }
                        ]
        };
```

TABLE 5.4 An anonymous *perl* hash referenced by the scalar VAR1 generated by the *XMLin* function of *XML::Simple* using the *Force Array* option.

```
The water molecule has 3 atoms
The methanol molecule has 6 atoms
```

We see that the first element of each array denoted by the index [0] is printed in the `foreach` loop.

5.2.5 Shapes

Consider the following *xml* data recorded in a file entitled *shapes.xml* containing elements described by text and attributes:

```
<?xml version="1.0" encoding="ISO-8859-1"?>
<shapes>
   <shape sid="2">rectangle</shape>
   <shape sid="1">triangle</shape>
   <shape sid="3">square</shape>
</shapes>
```

The attribute "sid" stands for shape identification number. The name of the root element is **shapes**. Running the following *perl* script residing in a file entitled *shapes.pl*:

```
#!/usr/bin/perl
#
use XML::Simple;      # use a perl module
use Data::Dumper;      # use a perl module
#
$utah= XMLin("shapes.xml");      # read XML file
print Dumper($utah);
```

prints on the screen the hash shown in Table 5.5(a). We see that, like elements, attributes are arranged in an anonymous hash array under the key `sid`. The content of each element expressing the shape type is indicated by a special key named `content`.

Appending to the script the lines:

```
foreach $item (0, 1, 2)
{
print "Shape id ", $utah->{shape}[$item]{sid}
    ," is a ", $utah->{shape}[$item]{content},"\n";
}
```

prints on the screen the following lines:

```
Shape id 2 is a rectangle
Shape id 1 is a triangle
Shape id 3 is a square
```

Key attribute

To arrange the shapes according to sid, we exercise the key attribute option by the line

```
$utah = XMLin("shapes.xml",KeyAttr=>"sid");
```

and obtain the hash shown in Table 5.5(b). We see that the depth of the *perl* array has been increased by one unit in response to the chosen attribute.

(*a*)

```
$VAR1 = {
          'shape' => [
                     {
                       'sid' => '2',
                       'content' => 'rectangle'
                     },
                     {
                       'sid' => '1',
                       'content' => 'triangle'
                     },
                     {
                       'sid' => '3',
                       'content' => 'square'
                     }
                   ]
        };
```

(*b*)

```
$VAR1 = {
          'shape' => {
                     '1' => {
                              'content' => 'triangle'
                            },
                     '3' => {
                              'content' => 'square'
                            },
                     '2' => {
                              'content' => 'rectangle'
                            }
                   }
        };
```

TABLE 5.5 Illustration of an anonymous *perl* hash referenced by the scalar VAR1 generated by the *XMLin* function of *XML::Simple* (*a*) without and (*b*) with the *key attribute* option.

Appending to the script the lines:

```
print "One ",$shapes_po->{shape}{1}{content}," found\n";
print "One ",$shapes_po->{shape}{2}{content}," found\n";
print "One ",$shapes_po->{shape}{3}{content}," found\n";
```

generates the following screen display:

```
One triangle found
One rectangle found
One square found
```

When a key attribute is used, the value of the corresponding element instead of its name is used as a key within the hash reference, serving as an index for accessing related data.

5.2.6 Converting perl structures to xml data

The *XMLin* function of *XML::Simple* discussed previously in this section arranges a given *xml* data structure into an appropriate *perl* structure. To accomplish the inverse, that is, convert a *perl* structure into a corresponding *xml* document, we use the *XMLout* function of *XML::Simple.*

As an example, we consider the *perl* code contained in the following file entitled *vivlia.pl*:

```
#!/usr/bin/perl
#
use XML::Simple;     # use a module
use Data::Dumper;     # use a module

$vivlia = {
    'mybooks' =>
    {
        'ncse' =>
          {
          'title' => 'Numerical Computation in Science & Engineering',
          'year' => '2008',
          'edition' => '2',
          'publisher' => 'OUP'
          },
        'fem' =>
          {
          'title' => 'Intro to Finite and Spectral Element Methods',
          'year' => '2005',
          'edition' => '1',
          'publisher' => 'CRC'
          },
    }
};

XMLout($vivlia,
        KeepRoot => 1,
        NoAttr => 1,
        XMLDecl => "<?xml version='1.0'?>",
        OutputFile => 'vivlia.xml',
        );
```

An anonymous *perl* hash referenced by the scalar `vivlia` is defined in the first part of the code. This hash contains only one key named `mybooks`, defining the root element of the corresponding *xml* document and representing an anonymous hash enclosed by the inner curly brackets (`{}`). The inner hash contains two keys representing books written by the author of the present book, each pointing to a hash where details are given regarding the book title, year of publication, edition, and publisher.

The function *XMLout* is called at the last section of the code with several self-explanatory options.

Running the *perl* script generates the following *xml* file entitled *vivlia.xml*, complete with *xml* declaration and a root element:

```
<?xml version='1.0'?>
<mybooks>
  <fem>
    <edition>1</edition>
    <publisher>CRC</publisher>
    <title>Intro to Finite and Spectral Element Methods</title>
    <year>2005</year>
  </fem>
  <ncse>
    <edition>2</edition>
    <publisher>OUP</publisher>
    <title>Numerical Computation in Science &amp; Engineering</title>
    <year>2008</year>
  </ncse>
</mybooks>
```

Exercises

5.2.1 *Install* perl

Install a version of *perl* on your computer and keep a log of the necessary procedures.

5.2.2 *Read an xml file*

Write a *perl* code that reads an *xml* file of your choice and prints selected elements.

5.2.3 *Print an xml file*

Write a *perl* code that prints an *xml* file contained data defined in a *perl* hash.

5.3 C++

The simplest method of generating *xml* output from C++ code is by hard-coding *print* statements for the individual *xml* elements, as discussed in Section

5.1 for *fortran*. The compilation and execution of C++ code was discussed in Section 1.6.3.

As an example, the following C++ code contained in a file entitled *exptab.cc* tabulates the exponential function and prints the results in a file entitled *exptab.xml*:

```
#include <iostream>
#include <fstream>
#include <iomanip>
#include <cmath>
using namespace std;

int main()
{
int i;
double step=0.1;

ofstream file1;
file1.open("exptab.xml");
file1<<setiosflags(ios::fixed | ios::showpoint);

file1<<"<?xml version='1.0' encoding='ISO-8859-1'?>" << endl;
file1<<"<?xml-stylesheet type='text/xsl' href='exptab.xsl'?>" << endl;
file1<<"<table>" << endl << endl;

for (i=1;i<=8;i++)
{
  double x=(i-1.0)*step;
  double y=exp(x);
  file1<<"<entry id='" << i <<"'>" << endl;
  file1<<"<x> " << setprecision(2) << setw(5) << x << "</x> ";
  file1<<"<y> " << setprecision(5) << setw(7) << y << "</y> " << endl;
  file1<<"</entry> " << endl;
}

file1<<"</table> " << endl;

file1.close();

return 0;
}
```

To compile this file in a *unix* system and create an executable binary file named *exptab*, we open a terminal and run the compiler by issuing the statement:

```
c++ -o exptab exptab.cc
```

and then pressing the ENTER key. Running the executable by typing:

```
./expatab
```

and then pressing the ENTER key produces the following file entitled *exptab.xml*:

```
<?xml version='1.0' encoding='ISO-8859-1'?>
<?xml-stylesheet type='text/xsl' href='exptab.xsl'?>
<table>

<entry id='1'>
  <x> 0.00</x> <y> 1.00000</y>
</entry>
<entry id='2'>
  <x> 0.10</x> <y> 1.10517</y>
</entry>
<entry id='3'>
  <x> 0.20</x> <y> 1.22140</y>
</entry>
<entry id='4'>
  <x> 0.30</x> <y> 1.34986</y>
</entry>
<entry id='5'>
  <x> 0.40</x> <y> 1.49182</y>
</entry>
<entry id='6'>
  <x> 0.50</x> <y> 1.64872</y>
</entry>
<entry id='7'>
  <x> 0.60</x> <y> 1.82212</y>
</entry>
<entry id='8'>
  <x> 0.70</x> <y> 2.01375</y>
</entry>

</table>
```

The name of the root element is `table`.

Next, we introduce the following *exptab.xsl* file referenced in the second line of the *xsl* file:

```
<?xml version="1.0" encoding="ISO-8859-1"?>
<xsl:stylesheet version="1.0"
    xmlns:xsl="http://www.w3.org/1999/XSL/Transform">

<xsl:template match="/">

<html>
  <body>
  <center>
```

Tabulation of the exponential

	x	exp(x)
1	0.00	1.00000
2	0.10	1.10517
3	0.20	1.22140
4	0.30	1.34986
5	0.40	1.49182
6	0.50	1.64872
7	0.60	1.82212
8	0.70	2.01375

FIGURE 5.3 Tabulation of the exponential generated by creating an *xml* file using a C++ code.

```
    <h2>Tabulation of the exponential</h2>
    <table border="2" cellspacing="4" cellpadding="4">
      <tr bgcolor="wheat">
        <th></th>
        <th >x</th>
        <th >exp(x)</th>
      </tr>
      <xsl:for-each select="table/entry">
      <tr bgcolor="ghostwhite" align="center">
        <td><xsl:value-of select="@id"/></td>
        <td><xsl:value-of select="x"/></td>
        <td><xsl:value-of select="y"/></td>
      </tr>
        </xsl:for-each>
      </table>
    </center>
    </body>
  </html>

  </xsl:template>
  </xsl:stylesheet>
```

When we open the *exptab.xml* file using a *web* browser, we obtain the display shown in Figure 5.3.

C++/xml libraries

Xml libraries are available for use with C++ code. *Libxml* is an *xml* parser and toolkit written in C, developed for the gnome project. A C++ wrapper is available.* The *Xerces-C++* library encapsulates functions that validate, parse, generate, and process an *xml* document using the document object model (DOM) or the simple application programming interface (API) for *xml* (SAX). Further information on these libraries can be found on the Internet.

Exercise

5.3.1 *Hard-coded xml*

Write a C++ code that produces *xml* formatted data of your choice in the output. Then write a companion *xsl* file that displays the data in the window of a *web* browser.

5.4 Matlab®

Matlab is an integrated application for interactive numerical computation and graphics visualization. The software was developed in the 1970s as a virtual laboratory for matrix calculus and linear algebra. Today, *Matlab* can be described both as a programming language and as a computational framework running on its own workspace inside an operating system empowering the hardware.

5.4.1 Generating xml data

Xml data can be generated by hard-coding *print* statements in a *Matlab* code, as discussed in Section 5.1 for *fortran* and in Section 5.3 for C++. The following *Matlab* script contained in a file named *eigen.m* computes the eigenvalues of an $N \times N$ matrix $\mathbf{A}$ using an internal *Matlab* function and prints the results in a file named *eigen.xml*:

```
N = 3;   % matrix size

A=[ 1 3 4;
  -3 5 6;
  -3 1 -4];

eigval = eig(A);

% now print

file = fopen('eigen.xml','w');

fprintf(file,'<?xml version="1.0" encoding="ISO-8859-1"?>\n');
```

*http://libxmlplusplus.sourceforge.net

```
fprintf(file,'<?xml-stylesheet type="text/xsl"href="eigen.xsl"?>\n');
fprintf(file,'<eigenvalues>\n\n');


for i=1:N
  fprintf(file,'<eigenvalue id="%2.0f">\n',i);
  fprintf(file,'\t<real>\n');
  fprintf(file, '\t\t%12.5f\n', real(eigval(i)));
  fprintf(file,'\t</real>\n');
  fprintf(file,'\t<imaginary>\n');
  fprintf(file, '\t\t%12.5f\n', imag(eigval(i)));
  fprintf(file,'\t</imaginary>\n');
  fprintf(file,'</eigenvalue>\n\n');
end

fprintf(file,'</eigenvalues>');
fclose(file);
```

Running this program produces the following file entitled *eigen.xml*:

```
<?xml version="1.0" encoding="ISO-8859-1"?>
<?xml-stylesheet type="text/xsl"href="eigen.xsl"?>
<eigenvalues>

<eigenvalue id=" 1">
  <real>
    -3.78985
  </real>
  <imaginary>
    0.00000
  </imaginary>
</eigenvalue>

<eigenvalue id=" 2">
  <real>
    2.89492
  </real>
  <imaginary>
    3.09226
  </imaginary>
</eigenvalue>

<eigenvalue id=" 3">
  <real>
    2.89492
  </real>
  <imaginary>
    -3.09226
  </imaginary>
```

```
</eigenvalue>

</eigenvalues>
```

The name of the root element is `eigenvalues`. The following *eigen.xsl* file referenced in the second line of the *xml* file is included in the same directory:

```
<xsl:stylesheet version="1.0"
    xmlns:xsl="http://www.w3.org/1999/XSL/Transform">
<xsl:template match="/">
<html>

  <body>
    <center>
    <h2>Matrix eigenvalues</h2>
    <table border="2" cellspacing="4" cellpadding="4">
    <tr bgcolor="wheat">
      <th>Eigenvalue</th>
      <th>Real part</th>
      <th>Imaginary part</th>
    </tr>
    <xsl:for-each select="eigenvalues/eigenvalue">
    <tr bgcolor="ghostwhite" align="center">
      <td><xsl:value-of select="@id"/></td>
      <td><xsl:value-of select="real"/></td>
      <td><xsl:value-of select="imaginary"/></td>
    </tr>
    </xsl:for-each>
   </table>
   </center>
 </body>

 </html>
 </xsl:template>
 </xsl:stylesheet>
```

When we open the *eigenvalues.xml* file using a *web* browser, we see the display shown in Figure 5.4. If we want to send the eigenvalues to another person, it is not necessary to also send pertinent explanations, as the *xml* data are self-explanatory.

Running xsl within Matlab

An *xml* file can be opened and processed according to instructions given in a companion *xsl* file inside the *Matlab* environment, thereby circumventing the need for a *web* browser or another *xsl* processor. In our example, at the *Matlab* prompt inside the *Matlab* environment, we issue the statement:

```
xslt eigen.xml eigen.xsl eigen.html -web
```

Matrix eigenvalues

Eigenvalue	Real part	Imaginary part
1	-3.78985	0.00000
2	2.89492	3.09226
3	2.89492	-3.09226

FIGURE 5.4 Eigenvalues displayed in a *web* browser based on an *xml* document generated from a *Matlab* code, subject to a companion *xsl* program.

which processes the file *eigen.xml* using the stylesheet *eigen.xsl* and writes the output to a file named *eigen.html*. The *html* file is then displayed in the *Matlab* help browser.

5.4.2 Using java to generate xml output

Java is an object-oriented compiled programming language similar to C++, but with a different object model. An attractive feature of *java* is that it handles memory allocation and deallocation efficiently in a way that is transparent to the programmer. A thorough understanding of *java* is not necessary for generating *xml* output from a *Matlab* code.

jvm

When a *java* code is compiled, a binary object file is produced in *bytecode*. This is not machine language, but rather an intermediate language that must be translated into machine language before execution. The translation is done by a program called the *java virtual machine* (*jvm*).

Java methods

Scientific programmers can identify a *java* method with a program function or subroutine. When the name of a method is encountered in a calling program, the method is executed. After the execution of the method has been concluded, data are transferred internally and control is passed to the calling program. *Java* is endowed with a wealth of methods arranged in different libraries.

Java and xml in Matlab

A *xml* library written in *java* and embedded in *Matlab* allows us to produce, import, and manipulate *xml* data. To read data from a *xml* document, we convert the *xml* document into a *java* object and extract its constituents. The use of *java* is mandated by the unavailability of native *xml* objects in *Matlab*.

Java in *Matlab* runs in its own workspace with a separate memory allocation. Data are transferred from the *Matlab* to the *java* workspace, as the need arises. This division of resources should be kept in mind when dealing with *xml* documents of large size.

Scientific programmers routinely use *Matlab* in a terminal (command-line window) by disabling the *java* virtual machine to reduce the memory requirements (*-nojvm* option). This option cannot be selected when *xml* manipulation is performed.

xmlwrite

In Section 5.4.1, we discussed an explicit method of generating an *xml* document by printing the individual elements using *Matlab*'s formatted output functions. As an alternative, we may use the following *java* method to map an *xml* document into a *java* object:

```
com.mathworks.xml.XMLUtils.createDocument('somename')
```

The method is illustrated in the following *Matlab* code contained in a file entitled *eiv.m*:

```
% Define a matrix

N = 3;

A=[ 1 3 4;
-3 5 6;
-3 1 -4];

% compute the eigenvalues

eigval = eig(A);

% create an xml Java object

dehesa = com.mathworks.xml.XMLUtils.createDocument('eigenvalues');
```

The last line creates a *java* object named `dehesa`, representing an *xml* document whose root element is named `eigenvalues`. The document is subsequently populated with elements according to the following code:

```
% run over the three eigenvalues

for i=1:N

  % create elements
```

```
    eg = dehesa.createElement('eigenvalue');    % eigenvalue
    eg_id = dehesa.createAttribute('id');
    eg_id.setNodeValue(sprintf('%2d',i));
    dehesa.getDocumentElement.appendChild(eg);
    eg.setAttributeNode(eg_id);

    re = dehesa.createElement('real');          % real part
    lamr = dehesa.createTextNode(sprintf('%5.3f',real(eigval(i))));
    eg.appendChild(re);
    re.appendChild(lamr);

    im = dehesa.createElement('imaginary');     % imaginary part
    lami = dehesa.createTextNode(sprintf('%5.3f',imag(eigval(i))));
    eg.appendChild(im);
    im.appendChild(lami);

end
```

The following *java* methods were used in this code:

```
createElement    createAttribute    createTextNode    setAttributeNode
setNodeValue     getDocumentElement.appendChild      appendChild
```

Finally, the *xmlwrite* function is used to print the *java* object using the line:

```
xmlfile = xmlwrite(dehesa)
```

Running the complete *Matlab* script consisting of the three components listed previously in this section prints on the screen the following lines:

```
xmlfile=

<?xml version="1.0" encoding="utf-8"?>
<eigenvalues>

  <eigenvalue id=" 1">
    <real>-3.790</real>
    <imaginary>0.000</imaginary>
  </eigenvalue>

  <eigenvalue id=" 2">
    <real>2.895</real>
    <imaginary>3.092</imaginary>
  </eigenvalue>

  <eigenvalue id=" 3">
    <real>2.895</real>
    <imaginary>-3.092</imaginary>
```

```
  </eigenvalue>

</eigenvalues>
```

A few empty lines have been added for visual clarity.

To save the *xml* data in a file named *eiv.xml*, we issue the following statement in the *Matlab* environment:

```
xmlwrite('eiv.xml',dehesa)
```

Perhaps ironically, hard-coding the *xml* print instructions, as discussed in Section 5.4.1, is more efficient than using *java* methods. This may not be the case in more involved applications.

5.4.3 Importing an xml document as a java object

We have seen that data can be arranged in an *xml* tree implemented as a *java* object. Conversely, an existing *xml* document can be imported into a *Matlab* session as a *java* object using the *xmlread* function.

For example, the *eigen.xml* document listed in Section 5.4.1 can be imported into a *Matlab* session as a *java* object arbitrarily named `eigen_jo` by the statement:

```
eigen_jo = xmlread('eigen.xml')
```

When printed on the screen by typing its name, the *java* object appears cryptically as:

```
eigen_jo = [ #document: null ]
```

The word `null` should not be alarming.

The *xml* document can be recreated from the *java* object using the *xmlwrite* function discussed in Section 5.4.2, by issuing the statement:

```
recreated = xmlwrite(eigen_jo)
```

where `recreated` is an arbitrary name. When printed on the screen, the recreated *Matlab* object appears as:

```
recreated =

<?xml version="1.0" encoding="utf-8"?>
<?xml-stylesheet type="text/xsl"href="eigenvalues.xsl"?>
<eigenvalues>
```

```
    <eigenvalue id=" 1">
    <real>
      -3.78985
    </real>
    <imaginary>
      0.00000
    </imaginary>
  </eigenvalue>

  <eigenvalue id=" 2">
    <real>
      2.89492
    </real>
    <imaginary>
      3.09226
    </imaginary>
  </eigenvalue>

  <eigenvalue id=" 3">
    <real>
      2.89492
    </real>
    <imaginary>
      -3.09226
    </imaginary>
  </eigenvalue>

</eigenvalues>
```

Some extra blank lines were deleted for clarity.

To save the recreated *xml* data in a file named *eval.xml*, we issue the *Matlab* statement:

```
xmlwrite('eval.xml',eigen_jo)
```

The complete *Matlab* script is contained in a file entitled *eval.m*.

We see that the combination of the *xmlread* and *xmlwrite* methods allows us to import and export *xml* documents, thanks to *java*.

5.4.4 Arranging xml data into Matlab structures

In a typical application, we are interested in importing *xml* data as input to code. Once imported as a *java* object, an *xml* document can be disassembled into desirable *Matlab* structures, such as vectors, matrices (arrays), and other objects, using suitable *java* methods. Some familiarity with *java* is necessary.

Data

As an example, we consider the results of an expensive experiment recorded in the following file entitled *data.xml*:

```
<?xml version="1.0" encoding="ISO-8859-1"?>
<?xml-stylesheet type="text/xsl"href="triangles.xsl"?>
<samothraki>

  <point>
    <x>0.0</x>
    <y>0.0</y>
  </point>

  <point>
    <x>1.0</x>
    <y>1.0</y>
  </point>

  <point>
    <x>2.0</x>
    <y>4.0</y>
  </point>

</samothraki>
```

The name of the root element of this *xml* document is `samothraki`. The following *Matlab* code residing in a file entitled *data.m* reads the data and produces the graph shown in Figure 5.5:

```
%=================================
% Read data from the data.xml file and prepare a graph
%
% data_jo:  data Java object
% enl:  element node list
% m:  length of enl
%=================================

data_jo = xmlread('data.xml');
enl = data_jo.getElementsByTagName('point');
m = enl.getLength();

for i=1:m

  % java array indices begin at 0

  ParentNode = enl.item(i-1);
  ParentNodeChildren = ParentNode.getChildNodes;
```

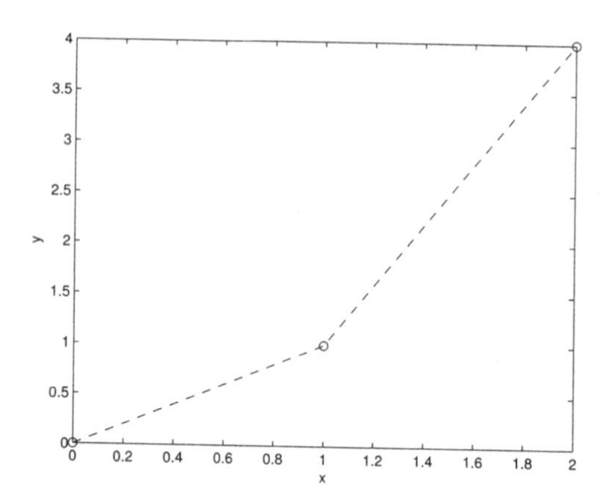

FIGURE 5.5 Graph of data read from an *xml* file using an imported Java object.

```
    % line breaks are counted as items:

    child1Node = parentNodeChildren.item(1);
    child2Node = parentNodeChildren.item(3);

    % get string values of items

    xs(i) = child1Node.getTextContent;
    ys(i) = child2Node.getTextContent;

    % convert to double precision

    x(i) = str2double(xs(i));
    y(i) = str2double(ys(i));

end

figure(1)
plot(x,y,'ko--');
xlabel('x');
ylabel('y')
```

The first line after the initial commentary employs the familiar *xmlread* method to read the data into a *java* object named `data_jo`. The second line generates a node list, arbitrarily named `enl`, of all points, regarded as children of `samothraki`. The third line extracts the size of the list recorded in the variable m using the `getlength` method. Running over the points in the ensuing loop allows us to arrange the x and y values of each point into two *Matlab* arrays.

```
Name                Size  Bytes  Class

child1Node           1x1        org.apache.xerces.dom.DeferredElementImpl
child2Node           1x1        org.apache.xerces.dom.DeferredElementImpl
data_jo              1x1        org.apache.xerces.dom.DeferredDocumentImpl
enl                  1x1        org.apache.xerces.dom.DeepNodeListImpl
i                    1x1     8  double
m                    1x1     8  double
parentNode           1x1        org.apache.xerces.dom.DeferredElementImpl
parentNodeChildren   1x1        org.apache.xerces.dom.DeferredElementImpl
x                    1x3    24  double
xs                   3x1        java.lang.String[]
y                    1x3    24  double
ys                   3x1        java.lang.String[]
```

TABLE 5.6 Miscellaneous variables employed in a *Matlab* code discussed in the text, containing a *java* object.

Note that line breaks between elements (nodes) in the *xml* file are counted as items of the corresponding *java* object, to be skipped when extracting textual content. This explains why the first and third entries of the array `parentnodeChildren` are selected in the fourth and fifth lines inside the *for* loop.

It is of interest to inspect the various variables introduced in this code. Issuing the *Matlab* statement:

```
whos
```

produces the list shown in Table 5.6, where *lang* stands for *language*. Variables that belong to the *org.apache.xerces* class cannot be viewed on the screen.

A few java methods

In the example discussed in this section, we have used a few *java* methods:

```
getElementsByTagName    getLength    getChildNodes    getTextContent
```

Additional useful methods are:

```
getTagName;    getFirstChild;    getLastChild;    getNextSibling
```

A summary of other methods accompanied by detailed explanations can be found at the Internet.*†

5.4.5 Navigating through an xml tree

Matlab procedures can be combined with *java* methods to navigate through an *xml* tree with the goal of identifying and extracting specific data of interest.

Polygon names

As an example, we consider data residing in a file entitled *polygons_inp.xml* containing the names of the first three regular polygons:

```
<?xml version="1.0" encoding="ISO-8859-1"?>
<?xml-stylesheet type="text/xsl"href="polygons.xsl"?>
<polygons>

  <polygon>
    <sides>3</sides>
    <name>equilateral triangle</name>
  </polygon>

  <polygon>
    <sides>4</sides>
    <name>square</name>
  </polygon>
  <polygon>
    <sides>5</sides>
    <name>pentagon</name>
  </polygon>

</polygons>
```

The name of the root element is `polygons`. The following *Matlab* code residing in a file entitled *polygons.m* returns the name of a polygon with a specified number of sides selected through the keyboard:

```
polygons_jo = xmlread('polygons.xml');

enl = polygons_jo.getElementsByTagName('polygon'); % element node list
m = enl.getLength();

n=200;   % an excessive number

% type the number of sides
```

*http://download.oracle.com/javase/6/docs/api/org/w3c/dom/Node.html
†http://download.oracle.com/javase/6/docs/api/org/w3c/dom/Element.html

```
while (n>5)
  n = input('Please enter the number of sides < 5:  ');
end

% run over the top-level xml elements inside the root element

for i=1:m

parentNode = enl.item(i-1);
parentNodeChildren = parentNode.getChildNodes;

% line breaks are counted as items:

sidesNode = parentNodeChildren.item(1);
nameNode = parentNodeChildren.item(3);

% get string values of items

test_sides = str2double(sidesNode.getTextContent);
test_name = nameNode.getTextContent;

if(n==test_sides)
  disp(test_name)
  break
end

end
```

A typical session follows:

```
Please enter the number of sides (less than 5): 3
equilateral triangle
```

5.4.6 Summary and toolboxes

We have seen that, thanks to *java*, *Matlab* is able to generate and read *xml* data. Several public *xml* toolboxes (add-on libraries) written by individual *Matlab* developers are available. For example, the *xml toolbox* converts *Matlab* data structures into *xml* trees. Most important, it also reads most types of *xml* trees converting them into appropriate *Matlab* structures.

Other proprietary computing environments, such as *Mathematica* produced by *Wolfram Research*, offer analogous facilities for importing, manipulating, and exporting *xml* data. Further information is given in tutorials available on the Internet.

Exercises

5.4.1 *xmlwrite*

Write a *Matlab* code that computes and prints in *xml* format all roots of a tenth-degree polynomial.

5.4.2 *xmlread*

Import an *xml* document containing the first five roots of the zeroth-order Bessel function, J_0, as a *java* object into *Matlab* code. Recreate and print the imported document.

5.4.3 *Indices*

Modify the *data.m* code listed in the text so that the index i runs from 0 to $m-1$.

5.4.4 *xyz plot*

Extend the code discussed in the text to generate a three-dimensional xyz graphs using data of your choice.

5.4.5 *Regular polyhedra*

Adapt the code for polygons discussed in the text to regular polyhedra.

5.4.6 *Xml toolboxes*

Discuss a *Matlab* *xml* toolbox of your choice and write a pertinent code.

A

ASCII code

Ascii is an acronym of the American Standard Code for Information Interchange. The *ascii* code maps 128 characters to an equal number of integers in the range 0–127 represented by the seven binary digits (bits). *Ascii* characters include letters of the English alphabet, digits, control characters, and other special symbols. The following general guidelines apply:

- Control characters for printers and other devices are encoded by the first 32 integers, 0–31. Code 32 represents the empty space between words.
- Codes 22–126 represent printable characters.
- The capital or upper-case letters of the English alphabet, A–Z, are encoded by successive integers in the range 65–90.
- The lower-case letters of the English alphabet, a–z, are encoded by successive integers in the range 97–122.
- The last code 127 is the Escape character.

The complete range of *ascii* characters is listed below:

Decimal	*Octal*	*Hex*	*Character*	
0	0	00	NUL	Null character
1	1	01	SOH	Start of header
2	2	02	STX	Start of text
3	3	03	ETX	End of text
4	4	04	EOT	End of transmission
5	5	05	ENQ	Enquiry
6	6	06	ACK	Acknowledgment
7	7	07	BEL	Bell
8	10	08	BS	Backspace
9	11	09	HT	Horizontal tab
10	12	0A	LF	Line feed
11	13	0B	VT	Vertical tab
12	14	0C	FF	Form feed
13	15	0D	CR	Carriage return
14	16	0E	SO	Shift out

15	17	0F	SI	Shift in
16	20	10	DLE	Data link escape
17	21	11	DC1	Device control 1 (usually XON)
18	22	12	DC2	Device control 2
19	23	13	DC3	Device control 3 (usually XOFF)
20	24	14	DC4	Device control 4
21	25	15	NAK	Negative acknowledgment
22	26	16	SYN	Synchronous idle
23	27	17	ETB	End of transmission block
24	30	18	CAN	Cancel
25	31	19	EM	End of medium
26	32	1A	SUB	Substitute
27	33	1B	ESC	Escape
28	34	1C	FS	File separator
29	35	1D	GS	Group separator
30	36	1E	RS	Record separator
31	37	1F	US	Unit separator
32	40	20	SPC	Space between words
33	41	21	!	
34	42	22	"	
35	43	23	#	
36	44	24	$	
37	45	25	%	
38	46	26	&	
39	47	27	'	
40	50	28	(	
41	51	29	)	
42	52	2A	*	
43	53	2B	+	
44	54	2C	,	
45	55	2D	-	
46	56	2E	.	
47	57	2F	/	
48	60	30	0	
49	61	31	1	
50	62	32	2	
51	63	33	3	
52	64	34	4	
53	65	35	5	
54	66	36	6	
55	67	37	7	
56	70	38	8	
57	71	39	9	
58	72	3A	:	
59	73	3B	;	

60	74	3C	<
61	75	3D	=
62	76	3E	>
63	77	3F	?
64	100	40	@
65	101	41	A
66	102	42	B
67	103	43	C
68	104	44	D
69	105	45	E
70	106	46	F
71	107	47	G
72	110	48	H
73	111	49	I
74	112	4A	J
75	113	4B	K
76	114	4C	L
77	115	4D	M
78	116	4E	N
79	117	4F	O
80	120	50	P
81	121	51	Q
82	122	52	R
83	123	53	S
84	124	54	T
85	125	55	U
86	126	56	V
87	127	57	W
88	130	58	X
89	131	59	Y
90	132	5A	Z
91	133	5B	[
92	134	5C	\
93	135	5D	]
94	136	5E	^
95	137	5F	_
96	140	60	`
97	141	61	a
98	142	62	b
99	143	63	c
100	144	64	d
101	145	65	e
102	146	66	f
103	147	67	g
104	150	68	h

105	151	69	i
106	152	6A	j
107	153	6B	k
108	154	6C	l
109	155	6D	m
110	156	6E	n
111	157	6F	o
112	160	70	p
113	161	71	q
114	162	72	r
115	163	73	s
116	164	74	t
117	165	75	u
118	166	76	v
119	167	77	w
120	170	78	x
121	171	79	y
122	172	7A	z
123	173	7B	{
124	174	7C	|
125	175	7D	}
126	176	7E	~
127	177	7F	DEL

B

Perl quick reference

Perl is a powerful interpreted programming language invented by Larry Wall, used in a wide variety of applications.* The term *interpreted* conveys that a *perl* program, typically called a *perl* script, does not have to be compiled into a binary executable. *Perl* is an acronym of the Practical Extraction and Reporting Language.

Locating the interpreter

The first statement in a *perl* script reveals the directory where the *perl* interpreter resides. In a standard *unix* system, this line reads:

```
#!/usr/bin/perl
```

Normally, the number sign (#) indicates a comment. However, the combination (#!), known as the *shebang*, is special.

To discover the location of the *perl* interpreter in a *unix* system, we may open a terminal (command-line window) and issue the command:

```
which perl
```

followed by the ENTER keystroke. The shell will reply with

```
/usr/bin/perl
```

Execution

To execute a *perl* script, we open a terminal and type the name of the file hosting the script followed by the ENTER keystroke,

```
./perlfile.pl
```

where *perlfile.pl* can be any file name. To ensure that the path of executables includes the current directory, we have inserted the dot-slash pair (./) in front of the *perl* file name.

*http://www.perl.com

Alternatively, we may issue the statement:

```
perl ./perlfile.pl
```

The *perl* file must be in the executable mode, which can be set using the *chmod unix* command.

Printing

The *print* function allows us to display character strings and numerical values on the screen. For example, we may state in a *perl* code:

```
print "avocado\n";
```

The character represented by the `\n` pair forces a line break. This statement prints the word `avocado` in a terminal, and moves the cursor to the next line.

To append text to the content of a file named *myfile.txt*, we issue the statements:

```
open (somename, '>> somefile.txt');
  print somename "this is an example\n";
close (somename)
```

Note that the double >> symbol is used. To overwrite existing information in a file, we use a single > symbol.

Scalar variables

The value of a scalar alphanumeric variable can be defined or extracted by prepending the dollar sign ($) to the variable name. For example, the following statements define and evaluate scalar variables:

```
$coefficient = 1.4;
$shape = "heptagon";
$somenumber= "345.6";
```

The dollar sign should be read as *value of. Perl* recognizes that the last variable holds a number, even though the numerical characters are enclosed by double quotes reserved for character strings. *Perl* variables are case sensitive and cannot begin with numbers.

The backslash indicates that the dollar sign is a literal (\$) to be treated as a character in a string, not as a scalar indicator. Thus, the statement:

```
print "This will cost you \$100.00 \n";
```

prints:

```
This will cost you $100.00
```

Arithmetic operations

The usual arithmetic operations can be performed between scalar variables holding numbers. Relational and logical operands for arithmetic operations are listed in Table 4.1. As in *fortran*, the power of a number is indicated by a double asterisk (**). Thus, the following line may appear in a *perl* script:

```
$power = $base ** $exponent;
```

The persistent use of the scalar *value of* operator ($) is a distinctive feature of *perl*.

String operations

Relational and logical operands for string manipulation are listed in Table 4.1. For example, the following code puts a hyphen between two words:

```
$a = "convection"; $b = "-"; $c = "diffusion";
$d = $a.$b.$c;
print "$d\n";
```

Running the script produces the output:

```
convection-diffusion
```

We deduce that the dot concatenates two character strings.

Arrays

The contents of an array are defined or extracted by prepending the *at* sign (@) to the array name. For example, the following statements define three *perl* arrays:

```
@prime_numbers = (1, 2, 3, 5);
@prime_numbers1 = (1, "two", 3, 5);
@primary_colors= ("red","green","blue");
```

The first scalar element of the array ***primary_colors*** defined in the third line is the scalar variable `primary_colors[0]`, the second scalar element is the scalar variable `primary_colors[1]`, and the third scalar element is the scalar variable `primary_colors[2]`. Note that array indices are enclosed by square brackets. The scalar #`primary_colors` is the index of the last element of the array.

To print the third element of the array `primary_colors`, we issue the statement:

```
print $primary_colors[2];
```

Issuing the statement:

```
print "$prime_numbers1[0] $prime_numbers1[1] $prime_numbers1[2] \n";
```

causes the following screen display:

```
1 two 3
```

The length of an array can be recorded in a scalar variable that is set equal to the array. In our example, we issue the statement:

```
$fountoukia = @primary_colors;
```

The scalar `fountoukia` is given the value of 3. Alternatively, the length of an array can be determined using the *scalar()* function.

Arrays can be used inside arrays, but the resulting data structure remains one-dimensional, described by a single index. For example, we may write:

```
@forgot_this = (5, 7);
@sequence = (1, 2, 3, @forgot_this, 11, 13);
print scalar(@sequence), "\n";
```

The last line prints the number 7.

Prepending a backslash (\) to the `@` sign indicates that the sign should be treated as a character literal.

Hashes

A *perl* hash, also called an associative array, allows us to link keys (variable names) to values (numbers or strings), as discussed in Sections 1.6.4 and 5.2. To evaluate or extract the contents of a hash, we prepend the percent sign (`%`) to the hash name. Curly brackets (`{}`) are used to access the value of a particular key.

As an example, the following complete *perl* code defines a hash and prints selected elements to convey a message:

```
#!/usr/bin/perl

%method = (
          context => "nonlinear equation",
          name => "Newton--Raphson",
          convergence => 2
          );
```

```
print "The $method{name} method is used to solve \n";
print "a $method{context} with convergence of order
    $method{convergence} \n";
```

Running the script produces the output:

```
The Newton–Raphson method is used to solve
a nonlinear equation with convergence of order 2
```

We observe that, like *xml* elements, *perl* hashes define and describe data, as discussed in Section 1.6.4.

References

A *perl* reference is a scalar variable representing another scalar, array, or hash. The reference is evaluated by prepending a backslash (\) to the represented scalar, array, or hash. The following line defines a reference to a string enclosed by double quotes:

```
$op = \"orthogonal polynomial";
```

A reference can be dereferenced by prepending to its name, including the prepended `$` sign, the `$` sign for a scalar or the `@` sign for an array. In our example, we may issue the statement

```
print $$op, "\n";
```

which prints

```
orthogonal polynomials
```

and moves the cursor to the next line.

The following statements define a *perl* array, assign a reference, print the array, and then print the array again in terms of the reference:

```
@prime_numbers = (1, 2, 3, 5);
$prime_numbers_reference = \@prime_numbers;
print "@prime_numbers\n";
print "@$prime_numbers_reference \n";
```

The output is:

```
1 2 3 5
1 2 3 5
```

for and for each

A `for` loop, also known as a `Do` loop, has the general structure:

```
for ($i=-34; $i<32; $i++)
{
  ...
}
```

A typical `foreach` loop has the general structure:

```
foreach $i (&somearray)
{
  ...
}
```

In both cases, the three dots indicate additional lines of code.

While and do while

A typical `while` loop has the general structure:

```
while (someexpression)
{
...
}
```

A typical `do while` loop has the general structure:

```
do
{
...
}
while (someexpression)
```

The loops are executed so long as `someexpression` is true.

if, else, unless

A typical `if` loop has the general structure:

```
if (someexpression)
{
...
}
```

A typical `if-else` loop has the general structure:

```
if (someexpression)
{
...
} else
{
...
}
```

A typical `if-elsif` loop has the general structure:

```
if (someexpression)
{
...
} elsif (anotherexpression)
{
...
}
```

A typical `unless` loop has the general structure:

```
unless (someexpression)
{
...
}
```

The loops are executed so long as `someexpression` is true.

Anonymous data

We may wish to reserve space in memory for anticipated data of a desired type that we are not prepared to name, but point to the data instead by a reference instead. To achieve this goal, we use anonymous data. Anonymous arrays are enclosed by square brackets (`[]`) instead of parentheses, and anonymous hashes are enclosed by curly brackets (`{}`) instead of parentheses. An arrow composed of two *ascii* characters (`->`) is used for dereferencing anonymous data.

As an example, we consider the following script:

```
@squares = (0, 1, 2, "four", "nine");      # named array
$gondor = \@squares;      # reference to a named array

$squirrel = [0, 1, 2, "four", "nine"];     # anonymous array

print "@$gondor[0] $squirrel->[1] @$gondor[2]
    $squirrel->[3] @$gondor[4] \n";
```

Running the script produces the output:

```
0 1 2 four nine
```

In this example, a named array and an anonymous array referenced by the variable `squirrel` are printed by reference.

An anonymous hash is defined in the following script:

```
$droplet = {
             fluid => "olive oil",
             diameter => "3 mm"
             };

print "Found a droplet of $droplet->{fluid}
    with diameter $droplet->{diameter}";
```

Running the script produces the output:

```
Found a droplet of olive oil with diameter 3 mm
```

Subroutines

Perl subroutines receive input and return scalar or array output. *Perl* variables are global, unless declared otherwise using the `my` qualifier in a subroutine or logical block. For example, we may define the local variable:

```
my $temperature = 37.1;
```

The following code prints the beginning of the Fibonacci series by calling a subroutine to add two numbers:

```
($n1, $n2) = (1, 1);
print "$n1 $n2 ";

for ($i=1; $i<16; $i++)
{
  $n3 = add_two_numbers();
  print "$n3 ";
  ($n1, $n2) = ($n2, $n3);
};

sub add_two_numbers{
  return ($n1 + $n2);
}
```

Running the script produces the output:

```
1 1 2 3 5 8 13 21 34 55 89 144 233 377 610 987 1597
```

The subroutine returns the scalar sum of two scalars communicated as global variables.

Exactly the same results can be generated by the alternative code:

```
($n1, $n2) = (1, 1);
print "$n1 $n2 ";

for ($i=1; $i<16; $i++)
{
  $n3 = add($n1, $n2);
  print "$n3 ";
  ($n1, $n2) = ($n2, $n3);
};

sub add{
return ($_[0] + $_[1]);
}
```

The subroutine `add` is called with two scalar arguments accommodated into an array named after the underscore (_).

Objects

Perl allows us to use objects, as discussed in *perl* language texts and manuals.

Summary of *xslt* elements C

In this appendix, we summarize available *xslt* 1.0 programming elements and illustrate their application. Elements that are deemed most useful in scientific computing are discussed in detail.

C.1 Stylesheet declaration, import, and inclusion

A stylesheet must be declared in the first line of an *xsl* file. Stylesheet import is available for splitting a code into multiple files.

- *stylesheet*

A stylesheet declaration at the beginning of an *xsl* file serves as a root element, to be closed at the end of the file. The general *stylesheet* declaration is:

```
<xsl:stylesheet version="1.0"
    xmlns:xsl="http://www.w3.org/1999/XSL/Transform"
    id="someid" exclude-result-prefixes="list"
    extension-element-prefixes="list" >
  ...
</xsl:stylesheet>
```

where the three dots indicate lines of *xml* code. The optional *id* identifies the stylesheet, the optional *exclude-result-prefixes* contains a list of namespace prefixes that will not be copied to the output, and the optional *extension-element-prefixes* contains a list of namespace prefixes used for extension elements.

Only the following eight *xslt* elements can be placed immediately after the *stylesheet* declaration:

```
xsl:template      xsl:attribute-set      xsl:import      xsl:include
xsl:output      xsl:param      xsl:script      xsl:variable
```

Additional stylesheets can be imported from other files using the *import* and *include* elements.

- *transform*

The *transform* declaration is synonymous to the *stylesheet* declaration. The general *transform* declaration is:

```
<xsl:transform version="1.0"
    xmlns:xsl="http://www.w3.org/1999/XSL/Transform"
    id="someid" exclude-result-prefixes="list"
    extension-element-prefixes="list" >
  ...
</xsl:stylesheet>
```

where the three dots indicate lines of *xml* code.

- *import*

The *import* element is used to import the contents of one stylesheet into another stylesheet according to the syntax:

```
<xsl:import href="anothersheet.xsl"/>
```

An imported stylesheet has lower precedence over the importing stylesheet.

- *apply-imports*

The *apply-imports* element is used to apply a template rule from an imported stylesheet according to the syntax:

```
<xsl:apply-imports/>
```

Rules stated in an imported stylesheet are overridden by those stated in the importing stylesheet. The *apply-imports* element can be used at any time to enforce an imported stylesheet.

- *include*

The *include* element is used to incorporate the contents of one stylesheet into another stylesheet according to the syntax:

```
<xsl:included href="anothersheet.xsl"/>
```

An included stylesheet has the same precedence as the including stylesheet.

C.2 Alternative code

If an *xslt* processor does not support a particular *xslt* element, alternative arrangements can be made.

- *fallback*

The *fallback* element encloses alternative code to be used if the *xslt* processor employed does not support an *xslt* element. A typical usage is:

```
<xsl:unrecognized_element>
    <xsl:fallback>
      ...
    </xsl:fallback>
</xsl:unrecognized_element>
```

where the three dots indicate lines of alternative (fallback) code.

C.3 Output formatting

An *xsl* code produces output that can be stored in a file or processed and displayed in a terminal or window of a *web* browser.

- *output*

The *output* element specifies the format of the output document according to the syntax:

```
<xsl:output
            method = "xml|html|text|name"
            version = "..."
            encoding = "..."
            omit-xml-declaration = "yes|no"
            standalone = "yes|no"
            doctype-public = "..."
            doctype-system = "..."
            cdata-section-elements = "..."
            indent="yes|no"
            media-type = "..."
            />
```

where the three dots denote a suitable string and a vertical bar denotes a choice. For example, unformatted data can be printed by specifying that `method="text"`. The default choice is `method="xml"`.

- *decimal-format*

The *decimal-format* element defines characters and symbols to be used when converting numbers into strings with the *format-number* function listed in Table D.1, Appendix D, according to the syntax:

```
<xsl:decimal-format
            name="decimal_format_name"
```

```
            decimal-separator="character"
            digit="character"
            grouping-separator="character"
            infinity="string"
            minus-sign="character"
            NaN="string"
            pattern-separator="character"
            percent="character"
            per-mille="character"
            zero-digit="character"
            />
```

The *decimal-format* element affects neither the *number* element, nor the *value-of* element, nor the *string* function.

- *number*

The *number* element is used to format a number or else determine the integer position of the current *xml* node in a list of nodes according to the syntax:

```
<xsl:number
            count="pattern"
            format="{ string }"
            from="pattern"
            grouping-separator="{ character }"
            grouping-size="{ number }"
            lang="{ languagecode }"
            letter-value={ "alphabetic|traditional" }
            level="any|multiple|single"
            value="expression"
            >
```

See also Section C.9.

- *namespace-alias*

The *namespace-alias* element replaces a namespace in the stylesheet with a different namespace in the output according to the syntax:

```
<xsl:namespace-alias stylesheet-prefix="unwanted_name"
    result-prefix="wanted_name" />
```

- *key*

The *key* element declares a named key that can be used in the stylesheet with the *key* function explained in Table D.3, Appendix D.

C.4 Comments, messages, and text

Comments can be inserted to explain symbols, insert clarifications, and announce important messages during execution.

- *comment*

The *comment* element is used to print a comment in the output enclosed between the standard `<!--` and `-->` delimiters according to the syntax:

```
<xsl:comment> this could be a clarification </xsl:comment>
```

- *message*

The *message* element is used to print a message and optionally terminate the processing of an *xml* document using a `yes` or `no` flag according to the syntax:

```
<xsl:message terminate="yes|no">
    The iterations did not converge.
</xsl:message>
```

Other *xsl* elements may be included inside the *message* element. In practice, the *message* element is used primarily to report errors.

- *text*

The *text* element is used to print text and optionally disable the interpretation of special characters such as the *greater than* (`>`) character aliased by the `>` string. A typical usage is:

```
<xsl:text disable-output-escaping="yes|no">
  The iterations did not converge.
</xsl:text>
```

Xsl elements may *not* be included inside the *text* element.

- *processing-instruction*

The *processing-instruction* element adds a processing instruction to the output. For example, the lines:

```
<xsl:processing-instruction name="xml-stylesheet">
  type="text/css" href="memo.css"
</xsl:processing-instruction>
```

produce the processing instruction:

```
<?xml-stylesheet type="text/css" href="memo.css" ?>
```

C.5 Xml element manipulation

An *xsl* code can modify the elements of an *xml* source document, thereby producing a new *xml* document with new, suppressed, or altered elements.

- *element*

The *element* programming element generates an *xml* element node at the output according to the syntax:

```
<xsl:element name="somename"
  namespace="URI" use-attribute-sets="somelist" >
  ...
</xsl:element>
```

where *URI* stands for a chosen universal resource identifier and the three lines indicate additional lines of code. The *namespace* and *use-attribute-sets* attributes are optional.

- *attribute*

The *attribute* element adds an attribute to an *xml* element according to the syntax:

```
<xsl:attribute name="somename" namespace="URI" >
  ...
</xsl:attribute>
```

where the three lines indicate additional lines of code. The *namespace* attribute is optional.

- *attribute-set*

The *attribute-set* element defines a named set of attributes according to the syntax:

```
<xsl:attribute-set name="somename" use-attribute-sets="somelist" >
  ...
</xsl:attribute-set>
```

where the three lines indicate additional lines of code. The *use-attribute-sets* attribute is optional.

- *copy*

The *copy* element creates a copy of the current node, omitting children nodes and attributes.

- *copy-of*

The *copy-of* element creates a copy of an *xml* node, including child nodes and attributes.

- *strip-space*

The *strip-space* element specifies a list of *xml* elements containing white (invisible) space alone to be removed from the output according to the syntax:

```
<xsl:strip-space elements="ISBN price" />
```

In this example, the ISBN number and the price of a book are not provided in the corresponding *xml* elements, and are therefore not included in the output.

- *preserve-space*

The *preserve-space* element reverses the action of the *strip-space* element according to the syntax:

```
<xsl:strip-space elements="ISBN edition" />
```

C.6 Logical constructs

Logical constructs are employed to repeat or selectively take an action, as discussed in Chapters 3 and 4.

- *if*

A typical usage of the logical *if* element is:

```
<xsl:if test="this_is_true or that_is_true">
  ...
</xsl:if>
```

where the three dots indicate additional lines of code. Logical `or` and `and` can be employed to combine tested conditions.

- *for-each*

The *for-each* element implements a repetition loop running over instances of a targeted *xml* node. A typical usage is:

```
<xsl:for-each select="xml_target_element">
  ...
</xsl:for-each>
```

where the three dots indicate additional lines of code.

• *choose, when, otherwise*

The *choose*, *when*, and *otherwise* elements implement conditional choices triggered by the veracity of tested expressions according to the syntax:

```
<xsl:choose>
  <xsl:when test="onecondition">
    ...
  </xsl:when>
  <xsl:otherwise>
    ...
  </xsl:otherwise>
</xsl:choose>
```

where the three dots indicate additional lines of code.

C.7 Variables and parameters

Local or global variables and parameters can be defined and evaluated in an *xsl* code by literal assignment or by using *xml* data, as discussed in Chapters 3 and 4. Global variables and parameters are declared before the root template of an *xsl* document, whereas local variables are declared inside the root template.

• *variable*

The *variable* element introduces and possibly evaluates a variable. One example is:

```
<xsl:variable name="radius">
  ...
</xsl:variable>
```

where the tree dots represent statements that evaluate the variable. A variable can be introduced and evaluated in a self-closing statement. Examples are:

```
<xsl:variable name="method" select="FFT"/>
<xsl:variable name="spectral_radius" select="1.1"/>
```

Further examples are discussed in Chapters 3 and 4.

• *param*

The *param* element introduces and possibly evaluates a parameter. Examples are:

```
<xsl:param name="radius">
  ...
</xsl:param>
```

where the tree dots represent statements that evaluate the parameter. Additional examples are:

```
<xsl:param name="method" select="GFEM"/>
<xsl:param name="accuracy" select="0.0001"/>
```

We see that a parameter can be introduced and evaluated in a self-closing statement.

- *with-param*

The *with-param* element introduces and possibly evaluates a parameter for use with a template according to the syntax:

```
<xsl:with-param name="somename" select="some_expression"/>
```

Examples are discussed in Chapters 3 and 4.

C.8 Templates

Templates are the counterparts of functions and subroutines. Matched, named, and matched and named templates are available. The use of templates is discussed extensively in Chapters 3 and 4.

- *template*

The *template* element defines a matched, named, or matched and named template. The declaration of a matched template is:

```
<xsl:template match="xml_node_name" priority="index" mode ="modename">
  <xsl:with-param name="somename" select="some_expression"/>
  ...
  <xsl:with-param name="lastname" select="last_expression"/>
    ......
</xsl:template>
```

where the three dots indicate additional *with-param* elements, and the six dots indicate additional code. The value of the ***priority*** attribute, ***index***, is a real number in the range $[-9.0, 9.0]$ with default value 0. The ***priority*** and ***mode*** attributes are optional. The ***mode*** attribute allows an *xml* node to be processed more than once. A matched template is called by the *apply-templates* element.

The declaration of a named template is:

```
<xsl:template name="template_name" >
  <xsl:with-param name="somename" select="some_expression"/>
  ...
  <xsl:with-param name="lastname" select="last_expression"/>
```

```
  ......
</xsl:template>
```

A named template is called by the *call-template* element.

The declaration of a matched and named template is:

```
<xsl:template match="node" name="somename"
    priority="index" mode ="modename" >
  <xsl:with-param name="somename" select="some_expression"/>
  ...
  <xsl:with-param name="lastname" select="last_expression"/>
    ......
</xsl:template>
```

A matched and named template is called either by the *apply-templates* element or by the *call-template* element.

- *apply-templates*

The *apply-templates* element applies all matched templates or a specific selected matched template according to the syntax: For example:

```
<xsl:apply-templates select="xmlnodename" mode="somemode" >
  <xsl:with-param name="somename" select="'some_text'"/>
  ...
  <xsl:with-param name="lastname" select="34.87"/>
  <xsl:sort select="some_element" order="ascending"/>
    ......
</xsl:apply-templates>
```

The *with-param* and *sort* elements inside the *apply-templates* element are optional.

- *call-template*

The *call-template* element calls for the execution of a named template according to the syntax:

```
<xsl:call-template name="somename">

  <xsl:with-param name="gac" select="9.81"/>
  ...
  <xsl:with-param name="name" select="'gravity'"/>

    ......
</xsl:apply-templates>
```

The *with-param* element inside the *call-template* element is optional.

C.9 Sorting and ranking

Xslt elements are available for sorting a specified list of numbers, characters, or strings in ascending or descending order, and for ranking the order of appearance of an element in a list or procedure.

- *sort*

The *sort* element sorts a homogeneous list consisting of entries with the same data type, including integers and character strings, in numerical or alphanumerical order according to the syntax:

```
<xsl:sort
select="something"
order= "ascending" or " descending"
data-type = "number" or "qname" or "text"
case-order = "upper-first" or "lower-first"
lang="language-key"
/>
```

All attributes are optional. The default order is ascending. The attribute `data-type = "qname"` is chosen for a user-defined data type.

- *number*

The *number* element determines the integer position of the current *xml* node in a list of nodes. Alternatively, the *number* element is used to format a number, as discussed in Section C.3.

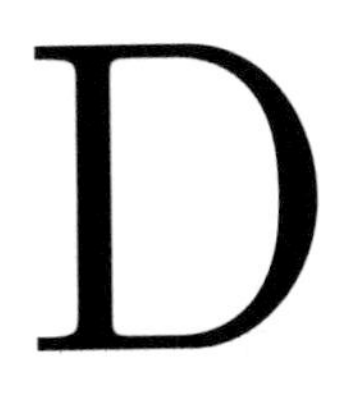

Functions called by xslt elements

Internal (core) *xslt* and *xpath* functions are available for evaluating the attributes of *xslt* programming elements, as discussed in Appendix C. Version 1.0 functions are implemented in standard *xsl* processors and *web* browsers. *Xslt* functions are shown in Table D.1, *xpath* functions of interest in numerical computation are shown in Table D.2, *xpath* functions of general interest are shown in Table D.3, and *xpath* functions for string manipulation are shown in Tables D.4 and D.5. Detailed explanations on the precise syntax and usage of these functions can be found in *xsl* texts and Internet tutorials.

Use of single quotes

It is important to note that string arguments of functions used to evaluate *xslt* element attributes are enclosed by single quotes ('), whereas the functions themselves are enclosed by double quotes (").

Add-on libraries

Add-on *xslt* function libraries are available. The *xslt* standard library (*xslsl*) contains a collection of templates written purely in *xslt*, including a limited number of mathematical templates useful in scientific computing.* The *exslt* library contains *xslt* templates implemented natively in *xslt* or in *javascript.*†

Xslt 2.0 functions

A broader set of functions is available in version 2.0 of *xslt* and *xpath.*‡ Selected functions are listed in Table D.6. Unfortunately, not all *xsl* processors and *web*browsers can process these functions. For example, *xslt* 2.0 libraries must be installed as an add-on in the Mozilla Firefox browser. The lack of extensive function libraries makes *xsl* attractive to scientific programmers who are challenged to demonstrate that a task can be accomplished with a minimum set of tools. From an educational perspective, *xsl* is an effective platform for teaching and developing programming skills.

*http://xsltsl.sourceforge.net

†http://www.exslt.org

‡http://www.w3.org/TR/xslt20

---

current () : Returns a node set that contains only the current node.

document ('uri_name', 'base_uri_name') : Discovers and parses an external *xml* document located at the specified uniform resource identifier (*uri*), and returns the node tree originating from the root element. The *base_uri_name* argument is optional.

element-available ('xsl:element_name') : Returns *true* or *false* according to whether an *xslt* element is supported by the *xslt* processor employed. For example, the statement:

```
<xsl:value-of select="element-available('xsl:sort')"/>
```

returns `true`, if the *sort* element is available. If an element is not available, "false" is returned and alternative arrangements must be made.

format-number (n, format, decimal_format_name) : Converts a number, `n`, into a string, and returns the string. If `n` is not a number, it is automatically converted into a number using the *xpath* `number()` function. The `format` argument appears as xxxxxx, where x can be # (number), dot (.), comma (,), semicolon (;), zero (0), or the percent character (%). The *decimal_format_name* argument is optional; see the *decimal_format xslt* element in Section C.3, Appendix C.

function-available ('function_name') : Returns *true* or *false* depending on whether an *xslt* or *xpath* function is supported by the *xslt* processor employed.

generate-id ('target_node') : Returns a character string serving as an *xml* node identification.

key ('node_name', 'node_value') : Returns a node set matching the name-value pair specified in the arguments.

TABLE D.1 *Xslt* 1.0 internal (core) functions to be used with *xslt* elements. Note the single quotes inside the parentheses holding the function arguments (*Continued*).

---

system-property ('information_requested') : Returns requested information on the *xslt* processor employed, such as *version* and *vendor*. For example, the lines:

```
XSL version
<xsl:value-of select="system-property('xsl:version')"/>
brought to you by
<xsl:value-of select="system-property('xsl:vendor')"/>
```

prints *"XSL version 1 brought to you by Transformiix"* in the current version of the Mozilla Firefox browser.

unparsed-entity-uri ('entity_name') : Returns the uniform resource identifier (*uri*) of an unparsed entity.

TABLE D.1 (*Continuing*) *Xslt* 1.0 internal (core) functions to be used with *xslt* elements. Note the single quotes inside the parentheses holding the function arguments.

---

number ('numstring') : Returns a number corresponding to a numerical string. If the argument `numstring` is omitted, the string content of the current node is assumed. For example, `number('-100.3')` returns `-100.3`, whereas `number('one_hundred')` returns `NaN` (not a number).

ceiling (number) : Returns the smallest integer that is greater than the number. For example, `ceiling(1.49)` returns `2`.

floor (number) : Returns the largest integer that is not greater than the number. For example, `ceiling(1.49)` returns `1`.

sum (target_node) : Returns the sum of the numeric values of a targeted node set. A typical usage is:

```
<xsl:value-of select="sum(element/child)"/>
```

count (target_node) : Returns the total element count of a targeted node set. A typical usage is:

```
<xsl:value-of select="count(element/child)"/>
```

true () : Returns the Boolean value *true.*

false () : Returns the Boolean value *false.*

boolean (argument) : Returns a Boolean value for a number, string, node set, or object. *True* is returned for a non-zero and not-a-number (NaN) number, a non-empty node-set, and a string with non-zero length. For example, `boolean(1)` returns *true.*

not (argument) : Returns a Boolean value for a number, string, or node set, or object. The argument is first reduced to a Boolean value by applying the *boolean* function, and then the negation condition is applied.

round (number) : Rounds a real number specified in the argument to the nearest integer. For example, `round(1.51)` returns `2`.

TABLE D.2 *Xpath* 1.0 functions of interest in scientific computing to be used with *xslt* 1.0 elements. Note the single quotes inside the parentheses holding the function arguments.

---

id ('id1 id2 ...') : Returns a set of nodes whose ID matches that specified in the argument.

lang (language) : Returns *true* if the language matches that of the context node, and false otherwise. The context node differs from the current node only when it is being tested for match.

last () : Returns the count of the last node in a list processed by the `xsl:for-each` or `xsl:apply-templates` element; see also the *position* function.

local-name (nodename) : Returns the local name of the qualified name of a node consisting of an optional prefix accompanied by a colon, and the node local name. If the argument is omitted, the name of the current *xml* context node is returned.

name (nodename) : Returns the name of a node consisting of an optional prefix accompanied by a colon, and the node local name. If the argument is omitted, the name of the current *xml* context node is returned.

namespace-uri (nodename) : Returns the node namespace *uri* of a node. If the argument is omitted, the name of the current *xml* context node is returned.

position () : Returns the rank of the current node in a list processed by the *xsl:for-each* or `xsl:apply-templates` element; see also the *last* function.

TABLE D.3 *Xpath* 1.0 functions of general interest to be used with *xslt* 1.0 elements. Note the single quotes inside the parentheses holding the function arguments.

---

concat ('item1', 'item2', ...) : Concatenates a list of character strings into a united string. For example the following statement returns *hello*:

```
<xsl:value-of select="concat('hell','o')"/>
```

contains (string, substring) : Returns true if the string contains the substring, and false otherwise. For example, the following statement returns *true*:

```
<xsl:value-of select="contains('hello','hell')"/>
```

normalize-space ('string') : Returns a string after removing extraneous blank space(s) prepended or appended to the string. If the argument is omitted, *string* is the name of the current *xml* context node. For example, the following statement returns *hello*:

```
<xsl:value-of select="normalize-space('hello')"/>
```

starts-with ('string', 'substring') : Returns true of *string* starts with *substring*, and false otherwise. For example, the following statement returns *true*:

```
<xsl:value-of select="starts-with('hello','hell')"
```

string ('somestring') : Returns the string equivalent of *somestring*, which could be a number. If the argument *somestring*, is omitted, the string equivalent of the current *xml* context node is returned.

string-length ('string') : Returns the number of characters in *string*. If the argument is omitted, the number or characters in the current *xml* context node is returned. For example, the following statement returns 5:

```
<xsl:value-of select="string-length('hello')"/>
```

substring ('string', n, m) : Returns a portion of *string* consisting of *m* characters after the *n*th character. The argument *m* is optional. For example, the following statement returns *genval*:

```
<xsl:value-of select="substring('eigenvalue',3,6)"/>
```

TABLE D.4 *Xpath* 1.0 functions for string manipulation to be used with *xslt* elements. Note the single quotes inside the parentheses holding the function arguments.

substring-after ('string','substring') : Returns the portion of a specified *string* following a specified *substring*. For example, the following statement returns *value*:

```
<xsl:value-of select="substring-after('eigenvalue','eigen')"/>
```

substring-before ('string', 'substring') : Returns the portion of a specified *string* preceding a specified *substring*. For example, the following statement returns *eigen*:

```
<xsl:value-of select="substring-before('eigenvalue','value')"/>
```

translate ('string1', 'string2', 'string3') : Returns a string that arises by replacing in *string1* the characters of *string2* with corresponding substitute characters contained in *string3*. For example, the following statement returns *EigEnValuE*:

```
<xsl:value-of select="translate('eigenvalue','ev','EV')"/>
```

TABLE D.5 *Xpath* 1.0 functions for string manipulation to be used with *xslt* elements. Note the single quotes inside the parentheses holding the function arguments.

---

abs (number) : Returns the absolute value of a number or numerical variable supplied in the argument. For example, `abs(-4.14)` returns `4.14`.

round-half-to-even (number, precision) : Receives a number and optionally a precision, and returns the number rounded to the nearest even integer with a specified precision. For example,

- `round-half-to-even(0.5)` returns 0
- `round-half-to-even(1.5)` returns 2
- `round-half-to-even(2.5)` returns 2
- `round-half-to-even(2.512, 1)` returns 2.5
- `round-half-to-even(2.512, 0)` returns 2

avg (sequence) : Returns the average of the argument values. For example, `avg((1,2,3))` returns 2.

max (sequence) : Returns the argument that is greater than all others in a sequence. For example, `max((1,2,3))` returns 3 and `max(('a', 'k'))` returns `k`.

min (sequence) : Returns the argument that is the lowest than all others in a sequence. For example, `min((1,2,3))` returns 1 and `min(('a', 'k'))` returns `a`.

TABLE D.6 *Xslt* 2.0 internal (core) functions useful for mathematical calculations. Not all *xsl* processors and *web* browsers support *xslt* 2.0 features and functions.

Index